Advancing Energy Policy

"The big transformations in a shift to a zero-carbon economy will be social and political, not only technological. This book provides engaging insights into the human dimensions of Europe's biggest energy policy challenges. Recommended reading for policy-shapers everywhere."
—Jonathan Gaventa, *Director, E3G, Belgium*

"This book provides compelling answers to important questions around energy-related Social Sciences and Humanities: why do we need it, how do we go about it and what is its impact? Both those committed to energy-SSH and those who are not (yet) will benefit greatly from the clear establishment of its necessity, actual workings and policy impacts. This makes this study likely to become a point of reference in the literature."
—Dr J.M. Wittmayer, *Senior Researcher, DRIFT, Erasmus University Rotterdam, The Netherlands*

"Builds a clear case for Social Sciences and Humanities as the missing link between energy related policy, practice and research."
—Dr Ruth Mourik, *DuneWorks, The Netherlands*

"Foulds and Robison have created an important resource for energy researchers, policymakers and practitioners. This powerful and informative edited volume offers guidance for those who want to understand the latest in the contributions of Social Sciences and Humanities to energy policy development."
—Professor Ramazan Sari, *Middle East Technical University, Turkey*

"Bring together energy researchers from the Social Sciences and Humanities, and the result is exciting. This is what think pieces really should be like. Ideas based on solid, interdisciplinary evidence leading to daring conclusions."
—Nils Borg, *Executive Director, European Council for an Energy Efficient Economy*

"An impressive take on contemporary energy policy issues with much needed fresh perspectives and an all-star roster of leading thinkers. I hope that every energy policymaker or even student of energy policy reads it."
—Benjamin K. Sovacool, *Professor of Energy Policy, University of Sussex, UK*

"The authors brilliantly demonstrate through a number of approaches, cases and examples, how interdisciplinary Social Sciences and Humanities research could and should be mobilised in EU energy policy and future energy transition research agendas."

—Marianne Ryghaug, *Professor of Science and Technology Studies, Norwegian University of Science and Technology*

Chris Foulds • Rosie Robison
Editors

Advancing Energy Policy

Lessons on the integration of Social Sciences and Humanities

Editors
Chris Foulds
Global Sustainability Institute
Anglia Ruskin University
Cambridge, UK

Rosie Robison
Global Sustainability Institute
Anglia Ruskin University
Cambridge, UK

ISBN 978-3-319-99096-5   ISBN 978-3-319-99097-2 (eBook)
https://doi.org/10.1007/978-3-319-99097-2

Library of Congress Control Number: 2018954415

© The Editor(s) (if applicable) and The Author(s) 2018. This book is an open access publication.
Open Access This book is licensed under the terms of the Creative Commons Attribution 4.0 International License (http://creativecommons.org/licenses/by/4.0/), which permits use, sharing, adaptation, distribution and reproduction in any medium or format, as long as you give appropriate credit to the original author(s) and the source, provide a link to the Creative Commons license and indicate if changes were made.
The images or other third party material in this book are included in the book's Creative Commons license, unless indicated otherwise in a credit line to the material. If material is not included in the book's Creative Commons license and your intended use is not permitted by statutory regulation or exceeds the permitted use, you will need to obtain permission directly from the copyright holder.
The use of general descriptive names, registered names, trademarks, service marks, etc. in this publication does not imply, even in the absence of a specific statement, that such names are exempt from the relevant protective laws and regulations and therefore free for general use.
The publisher, the authors and the editors are safe to assume that the advice and information in this book are believed to be true and accurate at the date of publication. Neither the publisher nor the authors or the editors give a warranty, express or implied, with respect to the material contained herein or for any errors or omissions that may have been made. The publisher remains neutral with regard to jurisdictional claims in published maps and institutional affiliations.

Cover illustration: © Melisa Hasan

This Palgrave Pivot imprint is published by the registered company Springer Nature Switzerland AG
The registered company address is: Gewerbestrasse 11, 6330 Cham, Switzerland

Foreword 1: Making Multiple Views Count—Why Energy Research Needs to Be Interdisciplinary

Gerd Schönwälder works on the socio-economic and political aspects of the clean-energy transition for the European Commission's Directorate-General for Research and Innovation (DG RTD). Previously, he was an invited researcher at the Centre for International Policy Studies (CIPS) and the German Development Institute (DIE), after holding senior positions at the International Development Research Centre (IDRC). Gerd earned a PhD in Political Science from McGill University.

Energy has always been political, but never more so than today. The transition to a cleaner, greener energy system profoundly affects not just individual lifestyles and livelihoods but entire societies, economies, even political systems. Prompting deep changes in the way people live, work and move around, the energy transition is generating innovative business models, novel ways to produce and deliver goods and services, as well as calls for greater involvement by consumers and citizens in relevant decision-making.

Energy *research*, by contrast, still mostly revolves around the *technical* challenges of moving from an energy system based largely on fossil fuels to one powered by renewables. The energy research landscape across Europe (and most of the world) remains fragmented, with insufficient exchanges between—as well as within—the Scientific, Technology, Engineering and Mathematics (STEM) disciplines on the one hand and the Social Sciences and Humanities (SSH) on the other. As a result, much-needed synergies that would require greater collaboration and more interdisciplinary work remain unrealised.

Initiatives such as SHAPE ENERGY want to change this. Supported by the European Commission's Horizon 2020 research framework programme, SHAPE ENERGY is narrowing the gap between Europe's energy research communities, reaching out to constituencies as varied as the business community, cities and Europe's citizens at large. In so doing, SHAPE ENERGY is contributing to the goals of the Energy Union and, more specifically, the Strategic Energy Technology Plan (SET-Plan), ensuring that Social Sciences and Humanities-related aspects have greater prominence in relevant energy research and energy policymaking. SHAPE ENERGY will lead to the establishment of a dedicated SSH platform alongside the existing energy European Technology and Innovation Platforms (ETIPs), starting in 2019.

The chapters in this collection make an important contribution to this agenda. They are stellar examples of the type of work that transcends not just disciplinary but also geographical boundaries, with the preparation of each chapter bringing together researchers from at least three SSH disciplines and two or more Horizon 2020 eligible countries. Transitioning to a cleaner-energy system, while building Europe's competitiveness and protecting its vulnerable citizens and regions, constitutes a fundamental challenge for the whole continent and such multiple perspectives are essential for confronting it. The contributions assembled here provide numerous insights that will be invaluable not just for researchers or policymakers but many others: cities, project developers, investors and of course concerned citizens all over the continent.

Gerd Schönwälder, European Commission (DG Research and Innovation [RTD])

Foreword 2: Multidisciplinary Partnerships for Access to Energy

Lidia Borrell-Damián has been Director for Research and Innovation (R&I) at the European University Association (EUA) since 2014, where she coordinates EU R&I project and policy development based on evidence provided by universities and National Rectors' Conferences. Areas of work include EU Programmes for R&I, EU Digital Agenda, Open Science, and Doctoral Education. In addition, she coordinates the EUA-Energy and Environment Platform (EUA-EPUE). She holds a Doctorate in Chemistry (photovoltaics) from the University of Barcelona.

The transition towards a carbon-neutral society or, preferably, towards a carbon-negative society requires the collective effort of all of us. It is now widely acknowledged that the Earth cannot sustain the pace at which its natural resources are being exploited and frequently converted into products that, even when they contribute to our well-being, are very difficult to reuse and recycle. At the bottom of the value chain for economic competitiveness and social prosperity lie the never-ending needs for affordable access to energy. Despite social inequalities and challenging political contexts, the world is slowly but surely solving the essential problems of access to water, food and health services (e.g. the rate of mortality in children under five has reduced by over 50% between 1990 and 2015). Now it is also time that our society reacts worldwide to provide more affordable access to clean energy to enable education and conditions for a hopeful future for all of us, while respecting our planet.

The Universities in the SET-Plan (UNI-SET) project (2014–2017) fostered a pan-European reflection on the role of universities in moving towards a 'cleaner-energy' society. More than 500 universities participated in dialogues over three years which led us to identify key areas of activity for the reform of educational programmes in energy efficiency, energy systems, renewable energy and many other domains of the energy field, producing the first 'Action Agenda for European Universities' for the energy transition. Very importantly, our reflection led us to realise that working towards our objectives requires in-depth interdisciplinary work and the integration of research approaches from Social Sciences and Humanities perspectives with those in Engineering and Natural Sciences. Moreover, integration of approaches *within* these two broad disciplinary areas are also necessary. For example, we need more civil engineers working with electrical engineers and with social scientists and humanists, in a true team effort to provide new ways to achieve sustainable access to energy in deprived areas, and solutions to save energy among those who enjoy a wealth of access to it. There is a vast amount of knowledge in great minds in our universities and research centres, and we need to bring them together within adequate partnership frameworks to further develop new knowledge that policymakers can use for the good of our society. Energy, environment and climate change issues are very closely related, and our current challenges need joint scientific and societal analyses to ensure that solutions are based upon the respect that nature and humanity as a whole deserve.

The collection of excellent chapters in this book, which arise from SHAPE ENERGY project activities, provide a series of valuable new insights and are examples of multidisciplinary thinking to tackle the energy transition. An underlying aspect in all these chapters, stated more or less explicitly, is the need to establish more and better partnerships, among experts in sciences, between experts and policymakers, between policymakers and citizens and so on. Ideas need to materialise into actions, which need to be governed by sound, honest and ethical principles; it is our planet that is at stake.

Lidia Borrell-Damián, Research & Innovation Unit, European University Association

Foreword 3: Energy Policies Outside the Silos

Ernst Ulrich von Weizsäcker has been Co-president of the Club of Rome since 2012. At the beginning of his career, Ernst served as professor and director of several universities and institutes. In 1991, he became founding President of the Wuppertal Institute. From 1998 to 2005, he was Member of the German Bundestag, chairing the Committees on Globalization and the Environment. He then served as Dean of the Graduate School of Environmental Science and Management at the University of California. In 2007, he was appointed Co-chair of the United Nations Environment Programme's (UNEP) International Resource Panel.

Current worldwide trends are not sustainable. The Club of Rome's warnings published in the book *Limits to Growth* in 1972 are still valid. We have, nevertheless, come a long way since 1972: we know much more about the climate and energy use and how what we do as societies affects the planet, for good and for bad. At times governments have been able to come together and effectively address threats to our survival, such as ozone-depleting gases. The United Nations Framework Convention on Climate Change's Paris Agreement is important, but far from sufficient, and now needs action to deliver upon the commitments made. We have also seen that energy use and economic prosperity can and have been decoupled, which is encouraging.

Still, current worldwide trends are not sustainable. Our societies still keep focusing on economic growth as the primary indicator for prosperity and while we reduce energy intensity, global energy use is still growing with potentially catastrophic consequences.

We seem to be in a philosophical crisis where our societies fail to address the problems that threaten their survival. The world needs a 'new enlightenment', one that is not based solely on doctrine but instead addresses a balance between humans and nature, as well as a balance between markets and the state and the short- versus long-term. To do this, we need to leave behind working in 'silos' in favour of a more systemic approach, which will require us to rethink the organisation of science and education.

This SHAPE ENERGY edited collection is therefore a timely publication and its interdisciplinary approach is especially encouraging. Researchers from around Europe firmly rooted in the Social Sciences and Humanities have produced ten texts that address energy issues from different angles. This is a refreshing departure from the common Economics- and Engineering-based approaches to 'solving' energy problems.

We need evidence-based knowledge in order to find solutions that work and are effective. This knowledge must be based on a deep understanding of the interaction between society and technology. The core chapters in this book offer insights into the socio-political characteristics of energy systems. They offer views on issues such as energy poverty, still often overlooked, but also expand into large-scale renewables deployment and the integration of electricity systems in Europe.

For policymakers who are used to looking at simple—perceived—causalities between investments and technology, these chapters are challenging in that they do not offer easy solutions. By reading and digesting these contributions, however, any person involved in energy policy and decisions about energy systems should find new perspectives and many eye-opening ideas to make him or her more prone to look for solutions that are based on an understanding of how people and our societies really work.

Ernst Ulrich von Weizsäcker, Club of Rome

Acknowledgements

We would like to thank all those who contributed pieces for this book, as well as several others for conversations and insights that fed into its design, including Hal Wilhite (University of Oslo), Nils Borg (European Council for an Energy Efficient Economy [eceee]), Mel Rohse (Anglia Ruskin University), Patrick Sumpf (Karlsruhe Institute of Technology), Christian Büscher (Karlsruhe Institute of Technology), Lauren Stabler (Anglia Ruskin University), Helga Hejny (Anglia Ruskin University) and the whole SHAPE ENERGY consortium.

The editors' time on this book project—and the collaboration expenses of chapter authors—was funded by the SHAPE ENERGY project, which is part of the European Union's Horizon 2020 research and innovation programme (under grant agreement number 731264). We also gratefully acknowledge Anglia Ruskin University's open access fund, which made the open access status of this publication possible.

We are thankful for internal Anglia Ruskin University support from Emma Milroy, Lenke Balint and Emma Rolph. Finally, we thank Rachael Ballard, Joanna O'Neill and Divya Anish of Palgrave Macmillan for their responsiveness and guidance throughout this process.

Contents

Foreword 1: Making Multiple Views Count—Why Energy
Research Needs to Be Interdisciplinary — v
Gerd Schönwälder

Foreword 2: Multidisciplinary Partnerships for Access to
Energy — vii
Lidia Borrell-Damián

Foreword 3: Energy Policies Outside the Silos — ix
Ernst Ulrich von Weizsäcker

1 Mobilising the Energy-Related Social Sciences
and Humanities — 1
Chris Foulds and Rosie Robison

Part I Energy as a Social Issue — 13

2 Plugging the Gap Between Energy Policy and the Lived Experience of Energy Poverty: Five Principles for a Multidisciplinary Approach — 15
Lucie Middlemiss, Ross Gillard, Victoria Pellicer, and Koen Straver

3 Shaping Blue Growth: Social Sciences at the Nexus Between Marine Renewables and Energy Policy — 31
Sandy Kerr, Laura Watts, Ruth Brennan, Rhys Howell, Marcello Graziano, Anne Marie O'Hagan, Dan van der Horst, Stephanie Weir, Glen Wright, and Brian Wynne

4 Looking for Perspectives! EU Energy Policy in Context — 47
Anna Åberg, Johanna Höffken, and Susanna Lidström

Part II Social Sciences and Humanities in Interdisciplinary Endeavours — 61

5 Achieving Data Synergy: The Socio-Technical Process of Handling Data — 63
Sarah Higginson, Marina Topouzi, Carlos Andrade-Cabrera, Ciara O'Dwyer, Sarah Darby, and Donal Finn

6 Building Governance and Energy Efficiency: Mapping the Interdisciplinary Challenge — 83
Frankie McCarthy, Susan Bright, and Tina Fawcett

7 Crossing Borders: Social Sciences and Humanities Perspectives on European Energy Systems Integration — 97
Antti Silvast, Ronan Bolton, Vincent Lagendijk, and Kacper Szulecki

8 A Complementary Understanding of Residential Energy
   Demand, Consumption and Services 111
   Ralitsa Hiteva, Matthew Ives, Margot Weijnen, and
   Igor Nikolic

Part III Interplay with Energy Policymaking Environments 129

9 Imaginaries and Practices: Learning from 'ENERGISE'
   About the Integration of Social Sciences with the EU
   Energy Union 131
   Audley Genus, Frances Fahy, Gary Goggins, Marfuga
   Iskandarova, and Senja Laakso

10 Challenges Ahead: Understanding, Assessing,
   Anticipating and Governing Foreseeable Societal Tensions
   to Support Accelerated Low-Carbon Transitions
   in Europe 145
   Bruno Turnheim, Joeri Wesseling, Bernhard Truffer, Harald
   Rohracher, Luis Carvalho, and Claudia Binder

11 Towards a Political Ecology of EU Energy Policy 163
   Gavin Bridge, Stefania Barca, Begüm Özkaynak, Ethemcan
   Turhan, and Ryan Wyeth

Afterword 1: Important Contributions Towards Renewal
of a Stubborn Energy Research and Policy Agenda 177
Harold Wilhite

Afterword 2: A New Energy Storyline 183
Inês Campos

Index 189

Notes on Contributors

Anna Åberg is Assistant Professor of the History of Science and Technology at the Chalmers University of Technology in Sweden. Her research areas include energy history and sports history.

Carlos Andrade-Cabrera is a doctoral candidate in Mechanical Engineering at University College Dublin.

Stefania Barca is Senior Researcher at the Center for Social Studies of the University of Coimbra. Her research interests intersect environmental history and political ecology.

Claudia Binder is Professor and Director of the Laboratory for Human-Environment Relations in Urban Systems, École Polytechnique Fédérale de Lausanne. Her research interests encompass analysing, modelling and assessing sustainability transitions.

Ronan Bolton works in Science and Technology Studies and Innovation Studies at the University of Edinburgh. His interests include the relationships between regulators, government, energy companies, users and local authorities.

Ruth Brennan is an interdisciplinary marine social scientist solicitor (non-practising), a Marie Skłodowska-Curie Individual Fellow at the Trinity Centre for Environmental Humanities, Trinity College Dublin, and an Honorary Fellow, Scottish Association for Marine Science.

Gavin Bridge is Professor of Geography at Durham University. His research centres on the political economy and political ecology of extractive industries and energy.

Susan Bright is Director of the Centre for Socio-Legal Studies at the University of Oxford, with a particular research interest in the challenges of upgrading multi-occupied buildings.

Luis Carvalho is Senior Researcher at the Centre of Studies in Geography and Spatial Planning at the University of Porto. His research deals with the geography of innovation and transitions.

Sarah Darby is Acting Leader of the Energy Programme at the University of Oxford's Environmental Change Institute, where she researches the social dimensions and environmental impacts of energy systems.

Frances Fahy is Senior Lecturer in Geography at the National University of Ireland (NUI), Galway and Lead Coordinator of the European Horizon 2020-funded ENERGISE project. Her research interests are in environmental planning and sustainability.

Tina Fawcett is Senior Researcher at the Environmental Change Institute, University of Oxford. Her research focuses on energy demand and energy policy.

Donal Finn is Associate Professor at the School of Mechanical & Materials Engineering, University College Dublin. His research interests include building energy systems and energy systems integration.

Chris Foulds is Senior Research Fellow at Anglia Ruskin University's Global Sustainability Institute and is co-lead of SHAPE ENERGY. His interests involve socio-technical change, energy demand, and policy interventions.

Audley Genus is Professor of Innovation at Kingston University and Work Package Leader on the European Horizon 2020-funded ENERGISE project. His research focuses on innovation and entrepreneurship for sustainability.

Ross Gillard is based at the Universities of Leeds and York. His research focuses on the social and political dimensions of sustainability, climate change and energy.

Gary Goggins holds a PhD in Environmental Sociology and Sustainability Studies and is the ENERGISE Project Manager at NUI Galway. His research interests are in sustainable consumption and knowledge transfer.

Marcello Graziano is Assistant Professor of Economic Geography at Central Michigan University and Member of the Institute for Great Lakes Research. His interests include regional economic modelling, energy and the energy-food-water nexus.

Sarah Higginson is an interdisciplinary researcher focusing on how timing and social practices influence domestic energy demand. She also designs processes that facilitate dialogue between stakeholders in sustainability issues.

Ralitsa Hiteva is Research Fellow in infrastructure governance, innovation and energy policy at the Science Policy Research Unit at the University of Sussex.

Johanna Höffken is Assistant Professor at the School of Innovation Sciences at Eindhoven University of Technology in the Netherlands. Her research focuses on energy and development in Asia.

Dan van der Horst is Reader in Energy, Environment and Society, at the School of Geosciences, University of Edinburgh. He works on environmental policy, energy transitions and societal change.

Rhys Howell is a postgraduate researcher in the School of Social and Political Science, University of Edinburgh. His current research focuses on the relationship between marine energy projects and communities.

Marfuga Iskandarova is a postdoctoral researcher with the ENERGISE project at Kingston University. Holding a PhD in Management Studies, her research interests include energy transitions, sustainable consumption and energy policy.

Matthew Ives is Senior Researcher at the University of Oxford's Institute of New Economic Thinking, where he researches transitions to a post-carbon society.

Sandy Kerr is an economist specialising in the blue economy, ocean governance, marine planning and renewable energy and Associate Professor and Director of the International Centre of Island Technology, Heriot-Watt University.

Senja Laakso is an environmental social scientist and postdoctoral researcher with the ENERGISE project at the University of Helsinki. Her research focuses on sustainable consumption, transformation of routines and social innovation.

Vincent Lagendijk is a historian of technology at Maastricht University. His publications include a book on the history of Europe's electrification and articles on blackouts, energy governance and international organisations.

Susanna Lidström is a researcher in Environmental Humanities at the Division of History of Science, Technology and Environment at KTH Royal Institute of Technology in Sweden.

Frankie McCarthy is Senior Lecturer in Law at the University of Glasgow. Her research centres around property law and property theory.

Lucie Middlemiss researches sustainable consumption and energy poverty in her position as Associate Professor in Sustainability, and Co-director of the Sustainability Research Institute, University of Leeds.

Igor Nikolic is Associate Professor at the Engineering Systems and Services Department of the Technology, Policy and Management Faculty at Delft University of Technology.

Ciara O'Dwyer is Senior Researcher at the School of Electrical and Electronic Engineering, University College Dublin. Her research interests include renewable generation integration, demand response and energy storage.

Anne Marie O'Hagan is Senior Research Fellow at the Centre for Marine and Renewable Energy Ireland, University College Cork, working on marine governance and environmental law.

Begüm Özkaynak is Professor at the Department of Economics, Boğaziçi University. Her research focuses on ecological distribution conflicts at the intersection of ecological economics and political ecology.

Victoria Pellicer researches energy poverty, sustainable transitions promoted by grassroots innovations and activism in citizen initiatives. She teaches on ethics and sustainability at the Universitat Politècnica de València.

Rosie Robison is Senior Research Fellow at Anglia Ruskin University's Global Sustainability Institute and co-lead of SHAPE ENERGY. She researches sustainable consumption, interdisciplinary working, 'smart' energy policy, and psychosocial interventions.

Harald Rohracher is Professor of Technology and Social Change at Linköping University. His research deals with the governance of sociotechnical change, infrastructure studies and the role of users in innovation.

Antti Silvast is a sociologist at Durham University working on energy social research, including energy systems integration. His monograph on electricity infrastructure was published by Routledge in 2017.

Koen Straver is a social psychologist at the Energy Research Centre of the Netherlands. He focuses on societal aspects and consequences of the energy transition, from local to global scales.

Kacper Szulecki is Assistant Professor of Political Science at the University of Oslo. He recently edited a book on energy securitisation, published by Palgrave, and a *Climate Policy* special issue.

Marina Topouzi is an interdisciplinary researcher in building energy use and demand, focussing on socio-technical factors in the 'performance gap' between intended/modelled design and actual performance of the built environment.

Bernhard Truffer is Professor at Utrecht University and Head of Environmental Social Sciences at the Swiss Federal Institute of Aquatic Science & Technology. He works on the geography of sustainability transitions.

Ethemcan Turhan is a postdoctoral researcher in the Environmental Humanities Lab at KTH Royal Institute of Technology in Stockholm. His main research interests are energy democracy and climate change politics.

Bruno Turnheim is Research Fellow at King's College London, the University of Manchester and the Laboratoire Interdisciplinaire Sciences Innovations Sociétés. His research focuses on innovation, sustainability transitions and their governance.

Laura Watts is an ethnographer of futures and Senior Lecturer in Energy and Society, University of Edinburgh. She is the author of *Energy at the End of the World: an Orkney Islands Saga*.

Margot Weijnen holds a Chair of Process and Energy Systems Engineering at the Department of Engineering Systems and Services at Delft University of Technology.

Stephanie Weir is a PhD candidate at the International Centre for Island Technology, Heriot-Watt University Orkney Campus, focussing on attitudes towards enclosure and privatisation in Scottish seas.

Joeri Wesseling is Assistant Professor at the Copernicus Institute of Sustainable Development, Utrecht University. He works on sustainability transitions from a socio-technical systems perspective.

Glen Wright is Research Fellow, Institute for Sustainable Development and International Relations, Paris, a PhD candidate, Australian National University and lead editor, *Ocean Energy: Governance Challenges for Wave and Tidal Stream Technologies*.

Ryan Wyeth is a PhD student in Geography at Durham University. His research focuses on the political economy and political ecology of hydroelectric development and water resources management.

Brian Wynne is Professor Emeritus of Science Studies and a former Research Director of the Centre for the Study of Environmental Change at Lancaster University.

LIST OF FIGURES

Fig. 5.1	Project description, design (top flowchart) and implementation (bottom flowchart)	65
Fig. 5.2	The timing of data collection across the two project strands	70
Fig. 5.3	Diagram of the subsystem integration and data flows behind the RealValue user interface application. WAN = Wireless Area Network, IoT = Internet of Things, SETS = Smart Electric Thermal Storage. Source: RealValue project partners, cited in Darby et al. (2018)	73
Fig. 5.4	Missing days of SETS monitoring data from Irish data sample ($n = 357$) in September 2017	75
Fig. 5.5	Sensor temperature data comparison in a single household in 2018 (potential misplacement)	76
Fig. 10.1	Share of energy from renewable sources in the EU Member States. Source: Eurostat (2018)	147

LIST OF TABLES

Table 5.1	Summary of data collected	68
Table 5.2	Actors involved in collecting and sharing different types of data sets	72
Table 6.1	Challenges of interdisciplinary research	92
Table 9.1	Comparing imaginaries: ENERGISE project proposal and H2020 SC3 (2014–15)	139
Table 10.1	Differences in formative and reconfiguration phase of energy systems change across multiple dimensions	150

LIST OF BOXES

Box 2.1	Netherlands vignette	19
Box 2.2	Spain vignette	19
Box 2.3	UK vignette	20
Box 3.1	Key opportunities for SSH-supported Marine Renewable Energy (MRE)	43
Box 4.1	Excerpt from the Citizens' summary of the Energy Roadmap 2050	49
Box 5.1	Attempts to collect temperature and occupancy data using technical and social methods	74
Box 7.1	Interdisciplinary workshop methodology for the development of research outputs	100
Box 8.1	The mythology of modelling and policy	112

CHAPTER 1

Mobilising the Energy-Related Social Sciences and Humanities

Chris Foulds and Rosie Robison

Abstract The energy-related Social Sciences and Humanities (energy-SSH) are commonly overlooked as a central evidence base for energy policy; the traditional Science, Technology, Engineering, and Mathematics (STEM) disciplines instead dominate the setting of policy goals. We argue that energy-SSH are insightful for energy policymaking and thus need more attention. We also make clear that to maximise their impact the considerable differences within energy-SSH need to be embraced rather than glossed over. From this position, we strongly advocate closer working of energy-SSH with STEM, as well as between the energy-SSH disciplines themselves. In illustrating all these points, we discuss the current European Union (EU) energy policy and research funding contexts and also outline our own SHAPE ENERGY project that aims to further the energy-SSH integration agenda across European circles. We finish the chapter with a brief commentary of this book's three core 'Parts', and their constituent chapters, which address different contributions and experiences of utilising energy-SSH.

Keywords Energy policy • Integration • Interdisciplinary • European Union • Horizon 2020 • SHAPE ENERGY

C. Foulds (✉) • R. Robison
Global Sustainability Institute, Anglia Ruskin University, Cambridge, UK
e-mail: chris.foulds@anglia.ac.uk; rosie.robison@anglia.ac.uk

© The Author(s) 2018
C. Foulds, R. Robison (eds.), *Advancing Energy Policy*,
https://doi.org/10.1007/978-3-319-99097-2_1

1.1 The Unfulfilled Potential of Social Sciences and Humanities in Driving (EU) Energy Policy

The range and significance of energy policy commitments made across local, regional, national, and international levels have been increasing. Many such commitments focus on guiding us through an 'energy transition' that entails energy system-wide changes aimed at various outcomes, be they regarding, for example, lower carbon emissions, increased security, interconnectedness, or affordability (Powell et al. 2015). The successful implementation of these policies and targets implies major changes for how energy is sourced, distributed, and consumed, with impacts for how all stakeholders (e.g. citizens, businesses, policymakers, other policyworkers, etc.) interact with the energy system on a variety of scales (Bridge et al. 2018; Walker and Cass 2007).

The European Union (EU) of course provides an excellent example of a framework within which such policy commitments are actively being made (European Commission 2017). At a strategic level, the EU is guided by its comprehensive integrated climate and energy policy,[1] which includes a number of 2030 targets: at least 40% reduction in emissions from 1990 levels, at least 27% supply from renewable energies, 27% (with a possibility of 30%) increase in energy efficiency, and cross-border interconnections for 15% of the EU's installed electricity production capacity. Alongside these headline targets, the EU has also constructed numerous policy frameworks, including flagship packages that include various policies within them (e.g. Clean Energy Package for All Europeans[2]), as well as more specific frameworks that are more targeted in their remit (e.g. Strategic Energy Technology Plan [SET-Plan][3]).

Alongside (and indeed sometimes in conjunction with) energy policy goals, there are commitments for new policymaking to be grounded in evidence. For example, the European Commission's Joint Research Centre (JRC), which is its in-house science advice service, has the core mission of providing 'EU policies with independent, *evidence-based* scientific and technical support throughout the whole policy cycle' (JRC in European Commission 2015, p. 5, emphasis added). Whilst we certainly acknowledge that there are debates around the merits/pitfalls of evidence-based policymaking (e.g. Pearce et al. 2014; Cairney 2016), including questions about the extent to which policies can be de-politicised and based on 'objective' and 'single-truth' evidence (Pielke Jr 2007; Robison and

Foulds 2018), we do nevertheless argue that it is vital to reflect on the role of 'epistemic communities'[4] which feed into evidence-gathering exercises and/or represent reference points for justifying energy policy positions. Moreover, we argue that Science, Technology, Engineering, and Mathematics (STEM) disciplines have dominated energy policy discourses in recent decades—including how society is or is not accounted for—as part of an established narrative of focusing on technological development (Guy and Shove 2000; Sovacool et al. 2015; Castree and Waitt 2017; Stirling 2014). Energy-related Social Sciences and Humanities (energy-SSH) disciplines are, in contrast, known to be commonly overlooked in favour of these technologically driven conventional alternatives (Foulds and Christensen 2016).

One aspect of this lack of involvement has arguably been that certain SSH approaches may be seen as representing all of SSH. Whilst bracketing energy-SSH together under the same umbrella term can be helpful in terms of building communities to promote the importance of socially grounded questions in energy, it is critical this does not come at the expense of neglecting the considerable variation within energy-SSH. As Fox et al. (2017, p. 3) note, 'energy-SSH' is not one homogenous mass of literature that is in (even approximate) agreement of how society is ordered; differences are everywhere'. We argue that such variation should be embraced, discussed frankly, and brought clearly to non-SSH audiences (including policy- or STEM-based groups), as opposed to imagining that a normalised, one-size-fits-all, homogenous version of SSH exists. This book showcases part of this variety.

Indeed, there are clear differences simply between the energy-related Social Sciences and the energy-related Humanities (Foulds et al. 2017; c.f. Castree et al. 2014, in terms of environmental-SSH). The energy-related Social Sciences (e.g. disciplines like Psychology, Sociology, Political Science) investigate the social organisation of human action, for example, attitudes, values, perceptions, norms, conventions, expectations, and so on, with an increasing interest in how these understandings could directly inform policy interventions. Whereas the energy-related Humanities (e.g. disciplines like History, Law, Theology) are concerned with the fundamental, and typically unspoken, cultural principles that underpin how societies are governed, for example, responsibilities, engagement, participation, (in)equality, equity, ethics, faith, and so on, with lessons for what societies should regard as 'desirable' (even if indirectly) when managing the energy system. There is great variation too within each of these disciplines and sub-disciplines concerning the theorisation and definition of the research problem in the first place (Sovacool and Hess 2017; Foulds and Robison 2017)—and this must not be forgotten.

As Hulme (2011, p. 178) states as part of his argument for overcoming the dominance of STEM and for embracing SSH difference (over consensus), particularly in terms of harnessing the potential of the Humanities: 'Crafting increasingly consensual reports of scientific knowledge, or levering more engineering and technology, will alone never open up pathways from research to the public imagination or the execution of policy'. Whilst his argument concerns climate science, there are inevitable parallels with the role of energy research in energy policy(making).

Discussions should therefore be developed with non-SSH energy research and policy communities on matters of SSH integration—that is, utilisation of key SSH concepts, understandings, methodologies, theoretical frameworks, and so on, in a way that meaningfully represents SSH on its own terms. However, we have found from our own experiences that energy policy-based advocates of 'interdisciplinarity' have for too long focused on how energy-SSH can support energy-STEM research, which has typically involved energy-STEM (and/or Economics, as one disputed discipline of the energy-related Social Sciences), setting the agenda for what role energy-SSH should play and thus how that disciplinary integration should be configured. We feel much more needs to be done to start new inclusive conversations on how energy-SSH could begin to take the strategic lead, through focusing more on energy-SSH in and of itself, and by exploring the potential of projects solely spanning insights from across energy-SSH.

In sum, the EU has (as indeed have other communities of policymakers) set significantly challenging commitments to change our energy system for the better. Such changes will inevitably need keen understandings of society's stakeholders, in terms of how and why they practically interact with all levels and elements of the energy system, as well as what the implications and consequences of those interactions may (or perhaps should) mean for society. Such evidence should be of real interest to those in policy circles and thus energy-SSH research needs to play more of a role somehow, and it is for this reason that this book aims to provide lessons on how energy-SSH should be recognised and better integrated into energy research and policy agendas.

1.2 Context: SHAPE ENERGY and the European Commission's Energy-Related Social Sciences and Humanities Work

A range of efforts are beginning to be made to undertake this integration work, on the ground. The European Commission's major research and innovation funding programme—Horizon 2020—uses the term 'Societal Challenges' to identify the areas of energy, transport, and so on, to which it allocates funding; however, it is the case that SSH expertise (i.e. which centrally considers *societal* processes and outcomes) is awarded a very much smaller proportion of this funding than STEM—4% vs. 96% of the €403M energy Work Programme budget in 2016, for example (European Commission 2018). Various initiatives have been designed to partially address this in recent years, including:

- a set of Horizon 2020 funding calls[5] explicitly for energy-SSH research, which has led to five dedicated energy-SSH projects being launched since 2016, with two of these represented in this collection (ENERGISE[6]; PROSEU[7]);
- a much larger number of energy topics being 'SSH-flagged'—that is, identified as needing SSH insights for their effective delivery (however, in 2016 almost 60% of these SSH-flagged energy calls included no partners with majority SSH expertise [European Commission 2018]); and
- a call to build a European Platform for energy-SSH[8] which could help bring its diverse communities together and, with a stronger voice, build its impact at a range of policy scales.

This latter call led to the creation of SHAPE ENERGY—Social sciences and Humanities for Advancing Policy in European Energy.[9] This Platform, which we designed and co-lead, began in February 2017 and has worked to develop Europe's expertise in using and applying energy-SSH. Specifically, we have worked to (1) understand and support interdisciplinary integration, (2) promote the role of energy-SSH to a range of stakeholders, and (3) gain greater insight into the needs of those who may wish to utilise energy-SSH, including policymakers. We have organised a range of activities aimed at different groups, often bringing stakeholders together across sectors, including running: academic and city-level multi-stakeholder workshops, a call for evidence, 'sandpits' for current Horizon 2020 consortia, PhD internships, and a Research Design Challenge, to name only a selection.

A significant activity in advancing our understanding of SSH integration for better energy policy has been commissioning this book, or series of 'think pieces', with colleagues from outside the SHAPE ENERGY consortium. This think piece book project has prioritised: (1) interdisciplinarity, and (2) collaboration. All pieces are co-authored by three or more researchers, with the discussions and research which fed into each individual chapter involving researchers from multiple European countries and three or more SSH disciplines. A competitive application process was run, with external peer reviewers, to identify which collaborations we should fund—this funding was then used for authors to meet and, in several cases, run events that fed directly into the chapters. The quality was high, which both meant we funded ten pieces rather than the planned eight and approached Palgrave Macmillan about publishing this collection as an open access book, not least because we felt that the contributions deserved wide exposure and would be of use to many.

We hope this book offers those new to SSH, or those interested in deepening their understanding across its span, a sense of the breadth and depth of what SSH can offer. This book will also be submitted to the energy strategy unit within the Directorate-General for Research and Innovation (DG RTD), as an official SHAPE ENERGY deliverable, to inform their ongoing work in the area. The EC has a particular interest in supporting the 'mainstreaming' approach whereby meaningful SSH involvement is recognised as needed to increase real-world impact of projects. Insights will also feed into SHAPE ENERGY's Research and Innovation Agenda 2020–2030, an output highlighting key challenges where energy-SSH can further provide direct leadership. But more broadly, we hope that others (e.g. away from the EC and following the completion of the SHAPE ENERGY project) will also be interested in reading the contributions in this book, including those working in practical energy initiatives aimed at furthering societal aims.

1.3 Structure of This Book

The Forewords, included prior to this Introduction, provide introductory remarks from three invited experts working in and around EU energy policy circles, who give their perspectives on the pressing energy challenges of our time and why SSH is needed to tackle these. The core of the book, from Chap. 2 onwards, then comprises ten short chapters from a total of 50 contributors. Each chapter is stand-alone and they can thus be read in any order.

The chapters are organised into three Parts, each of which reflects a commonly raised (and inevitably interconnected) theme regarding how energy-SSH can contribute to understanding of and/or working towards (EU) energy policy priorities. Thus, this book explores:

- *Part I.* Energy as a social issue;
- *Part II.* Social Sciences and Humanities in interdisciplinary endeavours; and
- *Part III.* Interplay with energy policymaking environments.

We briefly outline here the structure of these Parts to assist in navigating this collection.

Chapters 2, 3, and 4 in Part I set the scene through making the case for the centrality of social and human dimensions in the energy system. Whilst this is of course implicit throughout this book, these three chapters recognise that there is still a need to expose and highlight such dimensions; otherwise they risk being neglected. Each chapter clearly provides an answer to the request: 'explain to me why energy is a social issue'. Indeed, Middlemiss et al. (Chap. 2) begin with a powerful case for how qualitative understandings of the lived experience of energy poverty (gained through in-depth work with people) expose the limitations of narrow technical definitions, and can directly inform a more joined-up policy approach. Further, Kerr et al. (Chap. 3) take the case of an emerging technological field—the Marine Renewable Energy industry—and highlight the immediate and wide-ranging nature of the cultural, legal, and political issues surrounding so-called Blue Growth. Aberg et al. (Chap. 4) then take a very direct approach to illustrating how energy affects real lives—including raising issues of fairness and citizenship—through using three fictional stories from women across the world.

Given the social issues at stake, Chaps. 5, 6, 7, and 8 of Part II go on to explore the process of interdisciplinary working which seeks to involve SSH insights or methods, including in STEM-led projects. The chapters both discuss the challenges experienced but also, importantly, the impact interdisciplinary collaborations can have in responding to complex problems through making use of the latest understandings of the embedded relationships between technology and society. Higginson et al. (Chap. 5) provide a detailed and honest account of their journey through a collaboration that sought to bring together qualitative and quantitative data (on energy use in buildings); they provide several insights of use to cross-disciplinary endeavours. McCarthy et al. (Chap. 6) take the issue of building retrofits, often seen as an engineering problem, and discuss the need for understanding collective decision-making processes and legal frameworks when considering Multi-owned Properties. They recognise specific challenges for interdisciplinary research including when different SSH disciplines come together. Silvast et al. (Chap. 7) consider the concept of 'Energy Systems Integration (ESI)', which has emerged mainly from technical areas of research, and through discussion of concepts from History, Political Science, Sociology, and Science and Technology Studies, show how SSH can inform its development. Finally, in this Part, Hiteva et al. (Chap. 8) undertake a more theoretical analysis and comparison of different forms of modelling—techno-economic, agent-based, and ethnographic 'models'—which may be used in the policymaking process. They discuss both 'myths' that surround modelling, in addition to how different modelling approaches may be integrated together.

Building then on Hiteva et al.'s discussion of bringing disciplinary approaches together particularly for policy impact, in Part III, Chaps. 9, 10, and 11 consider the critical question of *how* understandings generated through SSH can be effectively brought to the policy table, and thus inform strategic planning. The very act of policymaking is a social process that SSH scholars have much interest in; energy-SSH does not merely concern itself with energy consumers, or 'end-users'. Genus et al. (Chap. 9) very directly confront the question of exactly how SSH integration is seen (or imagined) to have value in energy policy contexts currently, which then feeds into the shaping of funding calls. Turnheim et al. (Chap. 10) argue that Europe is now in an acceleration phase of renewables deployment, which raises fundamentally different questions both analytically and at a policy level than during earlier stages. (Energy-)SSH systems literature provides direct insights here both in identifying critical questions that need answering during this acceleration and in incorporating these into policy and practice. The final piece in this collection, Bridge et al. (Chap. 11), ends with a clear outline of the interdisciplinary field of political ecology, highlighting how its well-developed reflexive approaches can constructively challenge how policymaking is and should be done, in particular considering the role of social power in this process.

Taken as a whole, this book offers a window into the on-the-ground working of SSH in energy, and our two Afterword authors (Wilhite; Campos) extend this collection by offering their reflections on emergent themes, how the collection sits within the wider SSH literature, and what the work means for future energy-SSH projects and involvement at a European level. As editors, and through our experience of leading the SHAPE ENERGY Platform, we see real appetite to bring SSH better into the energy policy conversation. Notwithstanding the challenges that remain in implementing this, this book provides examples of how this is being and could be done.

NOTES

1. The comprehensive integrated climate and energy policy was adopted on 24 October 2014, as part of which there was a clear commitment to delivering the EU's 2030 targets (as detailed in the main text of this chapter). The subsequent 'Governance of the Energy Union' policy documentation ('COM(2016) 0759') was formally approved by the European Parliament on 30 November 2016; its purpose is to establish a framework to ensure those targets are achieved.
2. The 'Clean Energy Package for All Europeans' policy documentation ('COM(2016) 860') was formally approved by the European Parliament on 30 November 2016. Its purpose is to ensure that the EU remains competitive in the global energy market, mainly in response to anticipated changes associated with the clean energy transition. The Package includes eight different sets of legislative changes.
3. The Strategic Energy Technology Plan (SET-Plan) was adopted by the Commission on 22 November 2007. Its core purpose is to drive the development and diffusion of low-carbon/efficient energy technologies via strategically guiding the spending of research, development, and demonstration projects (primarily through its Horizon 2020 Framework Programme).
4. An epistemic community is 'a network of professionals with recognised expertise and competence in a particular domain and an authoritative claim to policy relevant knowledge within that domain or issue-area' (Haas 1992, p. 3).
5. EU Horizon 2020 LCE-31-2016-2017 funding calls, under the topic of 'Social Sciences and Humanities Support for the Energy Union'. Equivalent calls have also been released in the more recent energy Work Programme, specifically: EU Horizon 2020 LC-SC3-CC-1-2018-2019-2020, under the topic of 'Social Sciences and Humanities (SSH) aspects of the Clean-Energy Transition'.
6. Further details of the ENERGISE ('European network for research, good practice and innovation for sustainable energy') project are available at: www.energise-project.eu. A team from the ENERGISE consortium authored Chap. 9 of this book.

7. Further details of the PROSEU ('Prosumers for the Energy Union: mainstreaming active participation of citizens in the energy transition') project are available at: www.proseu.eu. The lead of PROSEU is the author of this book's second Afterword.
8. EU Horizon 2020 LCE-32-2016 funding call, under the topic of 'European Platform for energy-related Social Sciences and Humanities research'.
9. www.shapeenergy.eu.

References

Bridge, G., Barr, S., Bouzarovski, S., Bradshaw, M., Brown, E., Bulkeley, H., & Walker, G. (2018). *Energy and Society: A Critical Perspective*. Abingdon and New York: Routledge.

Cairney, P. (2016). *The Politics of Evidence-based Policy Making*. London: Palgrave Macmillan.

Castree, N., & Waitt, G. (2017). What Kind of Socio-Technical Research for What Sort of Influence on Energy Policy? *Energy Research & Social Science, 26*, 87–90.

Castree, N., Adams, W. M., Barry, J., Brockington, D., Büscher, B., Corbera, E., Demeritt, D., Duffy, R., Felt, U., Neves, K., Newell, P., Pellizzoni, L., Rigby, K., Robbins, P., Robin, L., Bird Rose, D., Ross, A., Schlosberg, D., Sörlin, S., West, P., Whitehead, M., & Wynne, B. (2014). Changing the Intellectual Climate. *Nature Climate Change, 4*, 763–768.

European Commission. (2015). *Strengthening Evidence Based Policy Making Through Scientific Advice—Reviewing Existing Practice and Setting Up a European Science Advice Mechanism*. Brussels: European Commission Directorate-General for Research and Innovation.

European Commission. (2017). *Third Report on the State of the Energy Union*. COM(2017) 688 Final. Brussels: European Commission.

European Commission. (2018). *Integration of Social Sciences and Humanities in Horizon 2020: Participants, Budget and Disciplines—3rd Monitoring Report on SSH Flagged Projects Funded in 2016 Under the Societal Challenges and Industrial Leadership Priorities*. Brussels: European Commission Directorate-General for Research and Innovation.

Foulds, C., & Christensen, T. H. (2016). Funding Pathways to a Low-carbon Transition. *Nature Energy, 1*(7), 1–4.

Foulds, C., & Robison, R. (2017). *The SHAPE ENERGY Lexicon—Interpreting Energy-related Social Sciences and Humanities Terminology*. Cambridge: SHAPE ENERGY.

Foulds, C., Fox, E., Robison, R., & Balint, L. (2017). *Editorial—Four Social Sciences and Humanities Cross-cutting Theme Reports*. Cambridge: SHAPE ENERGY.

Fox, E., Foulds, C., & Robison, R. (2017). *Energy & the Active Consumer—A Social Sciences and Humanities Cross-cutting Theme Report*. Cambridge: SHAPE ENERGY.

Guy, S., & Shove, E. (2000). *A Sociology of Energy, Buildings, and the Environment: Constructing Knowledge, Designing Practice*. London: Routledge.

Haas, P. M. (1992). Introduction: Epistemic Communities and International Policy Coordination. *International Organization, 46*(1), 1–35.

Hulme, M. (2011). Meet the Humanities. *Nature Climate Change, 1,* 177–179.

Pearce, W., Wesselink, A., & Colebatch, H. K. (2014). Evidence and Meaning in Policy Making. *Evidence & Policy, 10*(2), 161–165.

Pielke Jr., R. A. (2007). *The Honest Broker: Making Sense of Science in Policy and Politics*. Cambridge: Cambridge University Press.

Powell, J. C., Monahan, J., & Foulds, C. (2015). *Building futures: Energy Management in the Built Environment*. London: Routledge.

Robison, R. A. V., & Foulds, C. (2018). Constructing Policy Success for UK Energy Feedback. *Building Research and Information, 46*(3), 316–331.

Sovacool, B. K., & Hess, D. J. (2017). Ordering Theories: Typologies and Conceptual Frameworks for Sociotechnical Change. *Social Studies of Science, 47*(5), 703–750.

Sovacool, B. K., Ryan, S. E., Stern, P. C., Janda, K., Rochlin, G., Spreng, D., Pasqualetti, M. J., Wilhite, H., & Lutzenhiser, L. (2015). Integrating Social Science in Energy Research. *Energy Research & Social Science, 6,* 95–99.

Stirling, A. (2014). Transforming Power: Social Science and the Politics of Energy Choices. *Energy Research & Social Science, 1,* 83–95.

Walker, G., & Cass, N. (2007). Carbon Reduction, 'The Public' and Renewable Energy: Engaging with Socio-technical Configurations. *Area, 39*(4), 458–469.

Open Access This chapter is licensed under the terms of the Creative Commons Attribution 4.0 International License (http://creativecommons.org/licenses/by/4.0/), which permits use, sharing, adaptation, distribution and reproduction in any medium or format, as long as you give appropriate credit to the original author(s) and the source, provide a link to the Creative Commons license and indicate if changes were made.

The images or other third party material in this chapter are included in the chapter's Creative Commons license, unless indicated otherwise in a credit line to the material. If material is not included in the chapter's Creative Commons license and your intended use is not permitted by statutory regulation or exceeds the permitted use, you will need to obtain permission directly from the copyright holder.

PART I

Energy as a Social Issue

CHAPTER 2

Plugging the Gap Between Energy Policy and the Lived Experience of Energy Poverty: Five Principles for a Multidisciplinary Approach

Lucie Middlemiss, Ross Gillard, Victoria Pellicer, and Koen Straver

Abstract In this chapter, we illustrate the value of a multidisciplinary approach to energy poverty policy, drawing on insights from research into the lived experience of energy poverty in three European countries. We argue that understanding the lived experience of energy poverty is critical in designing energy policies which are fair, effective and

L. Middlemiss (✉) • R. Gillard
Sustainability Research Institute, University of Leeds, Leeds, UK
e-mail: L.K.Middlemiss@leeds.ac.uk; R.O.Gillard@leeds.ac.uk

V. Pellicer
Ingenio, Spanish Research Council (CSIC) & Universitat Politècnica de València, Valencia, Spain
e-mail: vicpelsi@ingenio.upv.es

K. Straver
Energy Transition Studies, ECN part of TNO, Amsterdam, Netherlands
e-mail: koen.straver@tno.nl

© The Author(s) 2018
C. Foulds, R. Robison (eds.), *Advancing Energy Policy*,
https://doi.org/10.1007/978-3-319-99097-2_2

aligned with people's daily lives. In addition, we contend that bringing together a range of disciplines to examine dimensions of the lived experience of energy poverty (such as housing, employment, education, social policy, health, energy, etc.) is essential to give breadth to our understanding of this challenging and multifaceted condition. We propose five principles for policy design, informed by our multidisciplinary understanding of the lived experience. These principles can be applied at a range of scales (local, regional, national and European) to help ensure that the energy poor are both well served, and represented, by energy policy.

Keywords Energy poverty • Energy vulnerability • Lived experience • Multidisciplinary

2.1 Introduction

Energy poverty is a fast-developing policy agenda at both European and other international levels. The launch of the European Union Energy Poverty Observatory (EPOV) in January 2018 marked an important moment in the connection of this policy agenda with academic research, as it is designed to encourage knowledge sharing and collaboration between policymakers, practitioners and academics in this field (EPOV 2018). It also reveals that the way different nations are driving this agenda is uneven: while policy on energy poverty is well established in some nations (the UK) and has made a strong start in others (Ireland, France), many nations around the European Union have yet to instigate policy on this topic. This policy agenda sometimes emerges at the local level (Spain and the Netherlands), in the absence of national targets or support (Straver et al. 2017). The agenda is sometimes resisted, or contested, with energy poverty being characterised as a problem of unemployment or poverty more generally (Germany, Spain, Denmark).

We are four energy poverty researchers, with a variety of disciplinary influences (Sociology, Social Policy, Psychology, Development Studies, Environmental Politics, Sustainability Social Science, Critical Geography and Policy Studies), committed to understanding the daily lives of energy poor households and to using that understanding to inform policy. The launch of EPOV, and the resulting attention being paid to the varied evolution of this agenda across the EU, gives us fresh impetus to argue for the importance of a multidisciplinary approach to energy poverty, and indeed energy policy more generally, based in a deep understanding of the lived experience.[1] Through our qualitative research and experiences in a number of European nations (the Netherlands, Spain, the UK), we have found that building a nuanced understanding of energy poverty, which takes into account the lived experience of fuel poor households, as well as how place and forms of vulnerability impact on those experiences, is essential in order to build meaningful policy and practice. In our work, we construct this broader picture by connecting research from disciplines active in researching the lived experience, with analysis of policy and practice on this topic. In doing so we make similar arguments to our colleagues writing in this volume about the importance of understanding daily life before attempting to intervene (Aberg et al., Chap. 4 in this collection). Here, we argue that integrating insights into the lived experience of energy poverty into policy and practice design is essential to ensure that action is meaningful and productive.

The growing body of academic research which aims to detail the lived experience of the energy poor (Day and Hitchings 2011; Middlemiss and Gillard 2015; Chard and Walker 2016; Butler and Sherriff 2017; Gillard et al. 2017; Pellicer-Sifres 2018) foregrounds a context-specific understanding of the varied challenges associated with a lack of access to energy services. Our own research on the lived experience of energy poverty in three European nations leads us to characterise this problem as multifaceted, and thus requiring a multidisciplinary response: it reaches into multiple domains of people's lives (housing, employment status, education, social relations, health, energy, etc.) and brings to light the interconnected nature of both these domains and the potential for vulnerability associated with these. These multiple dimensions of the problem, and the way they interact, are more likely to be unveiled by taking a multidisciplinary approach, drawing on lenses from different Social Sciences and Humanities disciplines.

As academics who research the lived experience, we frequently make alliances with practitioners whose work involves direct engagement with energy poor households. Based on these encounters with local activists, we are interested in ways of addressing the gap between the lived (local) experience and the design and delivery of policy interventions. Energy poverty policy aims to impact on people's daily lives, alleviating the challenges that they face and increasing their access to energy services. In approaching this problem through the lived experience, we notice that policy at the national level is failing to substantively address this problem on the ground (Middlemiss and Gillard 2015; Pellicer-Sifres 2018). In England, for instance, the measurement and definition of the problem of energy poverty creates a narrow interpretation, which does not reflect the complex and multifaceted nature of the lived experience (Middlemiss 2017). In our research in the Netherlands, local action and enthusiasm for this cross-cutting agenda has so far failed to stimulate a coordinated policy and investment schedule at a national level. Similarly, we find that in Spain, local policies willing to tackle energy poverty find resistance in national policies, which don't explicitly recognise the problem and therefore decline to modify laws and regulations. In each of these cases, a narrow understanding has produced technical and disconnected policy responses. Generally speaking, relying on just one or multiple aggregate indicators, such as income, demographic or geography, produces policies and schemes that are failing to meet the needs of households.

To remedy this, we call for a multidisciplinary approach that links the lived experience of the energy poor, to local, national or regional policy on the topic. To do that, we offer three vignettes (Boxes 2.1, 2.2 and 2.3) inspired by our empirical research in three different countries.[2] They show a range of life trajectories, allowing us to appreciate the complexity and the impact of different forms of vulnerability on the problem. In the vignettes, we show how energy poverty is linked to multiple dimensions of people's lives (housing quality, employment opportunities, health effects, etc.) and how existing policies either succeed or fail in tackling them. In Sect. 2.2, we reflect on the challenge of considering this complexity when designing and delivering policy, as well as the potential to address current policy shortcomings by interpreting these vignettes from a multidisciplinary perspective. In Sect. 2.3, we propose five principles for designing policy informed by the lived experience. These insights are also relevant to broader questions in energy policy about ensuring a fair transition to a low-carbon future, which we address in our conclusions.

Box 2.1 Netherlands vignette
Suzanne lives in Amersfoort with her two children Eva (6) and Mark (4). She was divorced last year and recently found herself in debt. She was left responsible for the mortgage, the costs of taking care of the children, groceries, and so on. Suzanne worked part time; her husband used to make a salary that covered most expenses. With her small salary she was not able to pay all monthly bills, and within five months her first reminders for payment turned into debts. There might be services, websites or municipality aids available to her, but she does not know where to find them or how to make use of them. The stress of taking care of the children and working as much as she can makes it difficult to find the time to fill in these forms. She has debts with her energy provider, amongst others. She does not know how to pay these debts, or how to save energy. To her, it is a fact of life and one of the many problems she's meaning to fix once things are less hectic.

There is no national policy on energy poverty in the Netherlands, which results in local governments that recognise this problem acting independently to tackle it. Therefore interventions for energy poverty are dispersed and temporary, with a common tendency to 're-invent the wheel', usually through short-term projects with low budgets. National data on the number of people struggling with energy poverty, or a coordinated national plan to support these people, are non-existent. From the perspective of the municipalities, housing corporations and health workers it is clear that helping households with energy advice can prevent debts, reduce expenditures, save energy, enhance living conditions and in some cases even create jobs when unemployed people are trained to give energy advice.

Box 2.2 Spain vignette
Tania and Manuel and their two daughters (three and five years old), based in Barcelona, have recently occupied an empty building owned by an important bank, with four other families. Tania works as a cleaner in an office, and Manuel has been working as a taxi driver for the last 30 years, but six months ago he was declared unable to work due to a health problem. Now, the family's income has been drastically reduced,

and they just manage to buy food and pay their water and energy bills, but they are unable to pay rent or any other extra expenses. Although they have paid their electricity bills, recently their energy company (one of the five biggest in Spain) cut their supply, arguing that they were living illegally. Fortunately, the family, together with neighbours, belongs to a social organisation fighting against energy poverty. Together they arranged a new electricity contract with a local citizens' energy cooperative, which did not ask them about their ownership. Tania and Manuel would not be able to negotiate this on their own, but bargaining collectively makes them feel safer.

In the city of Barcelona, the local municipality is trying to provide housing alternatives for families at risk of social exclusion, like Tania and Manuel. The council is negotiating with banks and private companies in order to make them rent out (at accessible and protected prices) some of the huge numbers of empty houses they have accumulated during the Spanish financial crises as a result of repossession of properties. Some of these empty houses are already occupied, but there is a lack of national regulation regarding when the energy company can cut off the supply in these situations: while the big five energy companies (considered to have political alliances with the banks who own these properties) reclaim ownership in order to supply energy, other small energy cooperatives recognise the problem of those families and offer facilities and discounts. The local municipality works closely with activists fighting against energy poverty, since they best understand the problems of local people.

Box 2.3 UK vignette
Clive is in his 50s and lives alone in an old terrace house in a small market town. After losing his job because of ill health, he was homeless for a long time. His house is rented, and it was the first one that the local council could offer him after being on a waiting list for many months. The house has draughty windows and doors, no wall or loft insulation and an inefficient heating system. Because he rents the property from the council, Clive has to wait for them to make any improvements to the

> house, because he can't afford to do them himself. Because Clive's ill health has left him almost immobile, he is not able to work regularly or get out much to socialise. He works 'cash in hand' jobs whenever he is well enough, but this income is not enough to pay all the bills, so he has stopped using the central heating and now only heats—and lives in— one room in the house. Because his work patterns and health are unpredictable, Clive doesn't claim state benefits or seek help with energy efficiency improvements—he never knows whether he is eligible or not and would struggle to find the necessary paperwork to prove it. Due to his social isolation, community health workers are the only people who see his living conditions, and nobody is aware of his precarious work and income situation.
>
> Social housing policy in the UK, at least where it is provided by local governments, is in such high demand that single adults without dependents have to wait a long time before they are eligible and have very limited choice. For someone like Clive, having to live in a poor-quality house in a relatively remote location is a major problem but it is his only option. Similarly, welfare support and energy efficiency policies are laden with conditions, leaving Clive confused and disinclined to investigate whether he is eligible for support. Ultimately, his current means of survival require him to work flexibly and cope with ill health almost on his own. Thankfully, the National Health Service in the UK provides community-based support, which means Clive gets to see health workers when he is ill. In this instance, there is an opportunity for the community health team to work across sectors and to provide Clive with additional advice and information, and to refer him to other means of support. Without this support, he would not receive the help he is entitled to.

2.2 A Multidisciplinary Approach to Plugging the Gap

Suzanne, Tania and Manuel, and Clive's stories show how vulnerabilities to energy poverty manifest themselves in a variety of ways. This reminds us of how complex an experience energy poverty is: it can intersect with challenges relating to health, social isolation, mobility, unemployment, education, housing, climate change, income, the energy market and energy regulations (and probably more). These intersecting dimensions result in different solutions being appropriate in different contexts. In addition, they make it essential to draw on the insights of multiple disciplines, from those painting a picture of the lived experience and beyond.

Drawing on a range of disciplinary insights to design and implement policy responses to energy poverty allows us to obtain a deeper appreciation of the causes and consequences of the problem, since it is likely to capture a more holistic description of people's experiences. For example, when a health researcher talks to someone experiencing energy poverty, they will elicit a different kind of response to a psychologist, a sociologist, an activist or a housing or poverty researcher (to name just a few). Where a health researcher might explore energy poverty consequences on physical health, a psychologist would focus on mental health, a sociologist would find difference regarding the social roles and power relations inside the household and an activist would be interested in empowering vulnerable people. When these disciplines are brought into conversation, they are likely to represent the experiences of the energy poor in a more nuanced and complete way.

In the world of policy and politics, the combined application of a number of disciplines could produce both practical recommendations and emotive arguments for addressing energy poverty. Arguably, politics and policy are two sides of the same coin, but productive action is more likely to be forthcoming when both are pulling in the same direction. With regard to the practicalities of policy: health, social care, energy and education tend to take an interest in households that are also vulnerable to energy poverty and could certainly share best-practice experiences. With regard to political agendas: energy poverty can provide emotive and powerful arguments for developing coalitions and drawing attention to injustices. For example, in the Netherlands, the NGO Milieudefensie financed research on the affordability of energy, which showed that Dutch industry gets more government subsidy for its energy use than Dutch households do, and, in addition, low-income households pay more for energy than high-income households do (Schep and Vergeer 2018). Combining policy insights from different domains on how to engage with energy poor households, together with political claims about distributional fairness, allows us to address this problem in a rather holistic way.

Combining disciplinary insights also has analytical benefits. For instance, creating a picture of the multi-dimensional nature of energy poverty through different disciplinary insights enables us to reveal the mismatches, overlaps and the unintended consequences of policies in different sectors. This is particularly important, given the complex nature of the unit of analysis (the household), at which the intersecting nature of many policies can be seen. In practice, engaging in multidisciplinary work on the lived experience also affords us opportunities to find ways of talking, and theorising, across disciplines. For instance, in our own work, we have used the concepts of energy justice, capabilities, social learning and social mobilisation to enrich our analysis and build collaborations with colleagues from different disciplines.

Such multidisciplinary and multi-sector work is often called for in public policy research and practice. For instance, 'policy integration' and 'joined-up service delivery' are common phrases in the literature, each stressing the potential benefits of cost savings, policy learning, multidisciplinary input, good governance, trust building and positive outcomes for the targets of policies (e.g. Entwistle and Martin 2005; Meijers and Stead 2004). Furthermore, valuing the lived experience and 'bottom up perspective' of practitioners is also a common feature in this literature. Research on distributional and procedural justice (Walker and Day 2012) within public policy makes a strong case for including the lived experience in all stages of the policy process: from agenda setting and policy formulation, right the way through to implementation and evaluation (Birkland 2015; Gillard et al. 2017).

2.3 Five Principles for Policy and Practice Informed by the Lived Experience

How might the understandings we can build from multidisciplinary work on the lived experience translate into policy at a national, subnational or supranational scale? In order to facilitate policy design which builds on the lived experience, we offer five guiding principles, each with a brief example evidencing their importance. These principles are based on our collective understanding of the possibilities for more integrated policy and practice, built on a combination of lived experience research, and thinking about the connection between multidisciplinary understandings and policy (see 'Acknowledgements' for a full account of the empirical work we are building on). These principles are intentionally broad and open to flexibility and future refinement, for example, there may be tensions between them and some may be more practicable than others depending on context and level of policymaking. These principles should also resonate beyond public administration, having relevance for energy companies, non-government organisations and charities who all encounter and work with energy poor households.

1. **Consider opportunities for joined-up and integrated policy**: A multidisciplinary understanding of the lived experience of energy poverty necessitates a clear commitment to coordinated action across multiple policy domains. As we have seen, it is frequently difficult to separate out policy domains and the impact they have on people's lives. For instance, people face health challenges as a result of the cold which can lead to, and be exacerbated by, unemployment, social isolation and deteriorating housing conditions. In the UK, there is a growing emphasis on the cross-over between health and energy poverty policy goals. As such, policy support is increasingly targeted at households with long-term health conditions, and partnerships with the health sector are being developed to help avoid costs to the healthcare system because of energy poverty.

   Given the privatised and liberalised nature of the energy industry in the UK, this also has to encompass non-state actors. Indeed, we see evidence of joined-up integrated action in practice when, for instance, private actors who service different domains attempt to coordinate their response to vulnerability (e.g. water, electricity and gas companies working together to share best-practice insights and to co-deliver support for vulnerable households such as the 'Stronger Together Coalition' in Wales).

2. **Building momentum through networks and partnerships**: The requirement for joined-up, integrated policy is always a challenge, given that governments and non-state actors might not have a history of working together on these matters. As a result, there is a need to build momentum through advocacy. This might include from below, such as in the Netherlands where the agenda is established at a local level, but less well recognised nationally. This can also be promoted by supply companies, as in the case of the Spanish citizen energy cooperative Som Energia, which has agreements with local councils from municipalities where Som Energia identifies defaults on bills. Once Som Energia identifies a household likely to

be energy vulnerable, it passes on this information to the department of social services in the city council. The cost of supply is shared by both Som Energia and social services, and an intervention on energy efficiency is led by local actors specialised in that domain. There are also opportunities to make alliances across nations, through transnational networks of local authorities and energy justice campaign groups, for instance. The current enthusiasm at the EU level is also helpful for this agenda, providing a top-down pressure for member states and policymakers to address the issue. Note that the purpose of advocacy here is to expose the multifaceted nature of this problem, and to engage a range of state and non-state actors in designing ways to address this problem which reflect the complexity of the lived experience.
3. **Expecting the unexpected**: Given that we know that this is a complex, and multifaceted, problem, policymakers and practitioners need to be alert to the possibility of intersecting challenges and unintended consequences. This requires flexible and reactive forms of governance, which incorporate opportunities for feedback, monitoring and evaluation. For example, many practitioners we have worked with report the co-occurrence of energy poverty and other social issues, such as mental health problems and social isolation, which presents unique challenges. Actors need to be equipped with the skills and resources to support households in the most appropriate ways. For example, service providers we have worked with in the UK noted that recognising—and responding to—the needs and expectations of energy poor households can sometimes require labour-intensive casework and 'bending the rules' of official policy frameworks, for example, around eligibility criteria. Often, organisations working on energy poverty find themselves dealing with complex mental health needs, helping households claim benefits they are entitled to or overlapping with social services in providing family support—all of which require significant amounts of personal skills, professional competence and resources.
4. **Measuring progress holistically**: Where national policy does exist, governments approach measurement of progress in a number of ways. Some governments are inclined towards a simple indicator for energy poverty (England), others opt for a 'basket' of indicators (France) and still others are more inclined towards an open approach

(Ireland). When informed by the lived experience, we argue that measurement must aim to capture the multifaceted nature of this problem: in practice this means drawing on multiple quantitative and qualitative indicators which relate to the various facets of energy poverty (e.g. income, housing, health, social isolation, mobility, climate change) to give a fuller picture of the problem and to allow unintended consequences to be observed. In addition, we should acknowledge the wider positive impacts of tackling energy poverty, such as improvements to infrastructure and housing, more community activity, local economic benefits and avoided costs to public services. In the city of Leeuwarden (the Netherlands), budget has been jointly allocated from the municipality and the province of Groningen for energy advisors to visit low-income households. The business case for this resource is made by summing up the saved CO_2 from the energy advice, the creation of jobs and the increase in income for households as a result of monthly savings.

5. **Just get on with it**: While our principles 1–4 suggest a slow and considered approach to this policy area, ensuring that we get to grips with its complexity and engage with its multiple facets, there is also much to be said for having a go and developing ideas through reflective practice. This might involve doing work in spite of the wider political and policy context, for example, advocating change and building capacity in local government networks and looking to international policy definitions and measurements to help build evidence. For example, in Spain, local government energy transition strategy often implies that deep changes in the political, economic and social arena are essential. This would mean removing power from dominant actors, and instigating public control. It would mean a clash with national government interests, which are focused on maintaining control of the energy market. Faced with this barrier, progressive initiatives must not stay paralysed, instead looking for alliances in new or powerful actors, such as the European Commission or a new body of energy cooperatives that have recently emerged. For instance, the three northern provinces of the Netherlands and 15 of their municipalities are currently producing an action plan to fight energy poverty in the region, as they see the multiple benefits of such a plan, and do not want to wait for national policy to be developed.

2.4 Conclusion

Our principles for policy and practice informed by the lived experience are important in thinking about energy poverty, but also have a role in influencing the broader direction of energy policy in relation to low-carbon futures. Energy policies increasingly attempt to incorporate justice and equity principles in their design, aiming for a fair and efficient transition towards a low-carbon energy system. An understanding of the lived experience of the energy poor, and how this is impacted by wider social and energy policy objectives, is essential in order to achieve an equitable future. In our research in relatively wealthy societies, people regularly have to make life- and health-limiting decisions about their access to energy. People's decisions are frequently based on trade-offs between different domains of their lives: maintaining good health, eating, heating and washing. Our own research, and others cited in this chapter, illustrate how qualitative research methods and context-sensitive engagement with households can enrich our understandings of lived experiences. For policymakers and frontline organisations, these methods can be utilised to inform policy development and evaluate its implementation.

Given that we already see substantial differences in people's lives depending on their access to these resources, there is a risk of the Matthew effect (Merton 1995: where rich become richer and poor become poorer) taking hold as we attempt to decarbonise energy supply. Indeed, if we are to achieve any kind of distributional justice in the future, building on lived experience research to avoid further deprivation for energy poor households is vital. The energy transition has the potential to increase living standards for all, but also holds the risk of further degrading the lives of the energy poor if policies are not integrated across domains and built on understandings of everyday life.

Acknowledgements The evidence base for this chapter is drawn from a range of research projects. The authors would like to acknowledge: the UK Energy Research Centre, Carolyn Snell and Mark Bevan from the University of York; the Spanish Ministry of Economy and Competitiveness [grant number: CSO2013-41985-R]; the knowledge platform on socially responsible innovation (the NWO-MVI Platform)—intended for science, business and social organisations; the White Rose Collaboration Fund; and colleagues working on Energy Poverty and Social Relations.

Notes

1. By 'lived experience', we mean qualitative, deep understandings of the daily lives of people who are categorised as experiencing energy poverty.
2. We built these vignettes at a workshop, following reflections on how lived experience research reveals the absence of adequate policy. The vignettes are based on real-life examples but are amalgamated characters designed to show the links between policy and everyday life.

References

Birkland, T. A. (2015). *An Introduction to the Policy process: Theories, Concepts, and Models of Public Policy Making*. Abingdon: Routledge.

Butler, D., & Sherriff, G. (2017). 'It's Normal to Have Damp': Using a Qualitative Psychological Approach to Analyse the Lived Experience of Energy Vulnerability Among Young Adult Households. *Indoor and Built Environment, 26*(7), 964–979.

Chard, R., & Walker, G. (2016). Living with Fuel Poverty in Older Age: Coping Strategies and Their Problematic Implications. *Energy Research & Social Science, 18*, 62–70.

Day, R., & Hitchings, R. (2011). 'Only Old Ladies Would Do That': Age Stigma and Older People's Strategies for Dealing with Winter Cold. *Health & Place, 17*(4), 885–894.

Entwistle, T., & Martin, S. (2005). From Competition to Collaboration in Public Service Delivery: A New Agenda for Research. *Public Administration, 83*(1), 233–242.

European Energy Poverty Observatory. (2018). European Energy Poverty Observatory Website [online]. Retrieved March 5, 2018, from https://www.energypoverty.eu/

Gillard, R., Snell, C., & Bevan, M. (2017). Advancing an Energy Justice Perspective of Fuel Poverty: Household Vulnerability and Domestic Retrofit Policy in the United Kingdom. *Energy Research & Social Science, 29*, 53–61.

Meijers, E., & Stead, D. (2004). Policy Integration: What Does It Mean and How Can It Be Achieved? A Multidisciplinary Review. Conference proceedings of *2004 Berlin Conference on the Human Dimensions of Global Environmental Change: Greening of Policies—Interlinkages and Policy Integration*. 3–4 December 2004, Berlin, Germany.

Merton, R. K. (1995). The Thomas Theorem and the Matthew Effect. *Social Forces, 74*(2), 379–422.

Middlemiss, L. (2017). A Critical Analysis of the New Politics of Fuel Poverty in England. *Critical Social Policy, 37*, 425–443.

Middlemiss, L., & Gillard, R. (2015). Fuel Poverty from the Bottom-up: Characterising Household Energy Vulnerability Through the Lived Experience of the Fuel Poor. *Energy Research & Social Science, 6*, 146–154.

Pellicer-Sifres, V. (2018). Ampliando la comprensión de la pobreza energética desde el enfoque de capacidades: hacia una mirada construida desde las personas afectadas. *Iberoamerican Journal of Development Studies.* Forthcoming. Retrieved April 15, 2018, from http://ried.unizar.es/public/abstracts/ampliandolacomprension.pdf

Schep, E., & Vergeer, R. (2018). Indicators for a Just Climate Policy. 18.7N38.02a. https://www.ce.nl/publicaties/2061/indicatoren-voor-een-rechtvaardig-klimaatbeleid

Straver, K., Siebinga, A., Mastop, J., De Lidth, M., Vethman, P., & Uyterlinde, M. (2017). Effective Interventions to Enhance Energy Efficiency and Tackle Energy Poverty. ECN-E--17-002. https://www.ecn.nl/publicaties/PdfFetch.aspx?nr=ECN-E%2D%2D17-002

Walker, G., & Day, R. (2012). Fuel Poverty as Injustice: Integrating Distribution, Recognition and Procedure in the Struggle for Affordable Warmth. *Energy Policy, 49*, 69–75.

Open Access This chapter is licensed under the terms of the Creative Commons Attribution 4.0 International License (http://creativecommons.org/licenses/by/4.0/), which permits use, sharing, adaptation, distribution and reproduction in any medium or format, as long as you give appropriate credit to the original author(s) and the source, provide a link to the Creative Commons license and indicate if changes were made.

The images or other third party material in this chapter are included in the chapter's Creative Commons license, unless indicated otherwise in a credit line to the material. If material is not included in the chapter's Creative Commons license and your intended use is not permitted by statutory regulation or exceeds the permitted use, you will need to obtain permission directly from the copyright holder.

CHAPTER 3

Shaping Blue Growth: Social Sciences at the Nexus Between Marine Renewables and Energy Policy

Sandy Kerr, Laura Watts, Ruth Brennan, Rhys Howell, Marcello Graziano, Anne Marie O'Hagan, Dan van der Horst, Stephanie Weir, Glen Wright, and Brian Wynne

Abstract The development of the Marine Renewable Energy (MRE) industry is part of the EC Blue Growth Strategy. It brings together a range of relationships across people, sea, and energy, from developers to local communities and policymakers. This calls for diverse approaches, moving beyond an oppositional mindset to one that can establish an inclusive community around MRE development. Ownership of the marine environment is a legal issue, but MRE devices operate within a cultural and emotional

---

S. Kerr (✉) • S. Weir
International Centre of Island Technology, Heriot-Watt University, Edinburgh, UK
e-mail: s.kerr@hw.ac.uk; Sw34@hw.ac.uk

L. Watts (✉)
Institute of Geography and the Lived Environment, Edinburgh University, Edinburgh, UK

Technologies in Practice, IT University of Copenhagen, Copenhagen, Denmark
e-mail: l.watts@ed.ac.uk

sense of place. Early, sustained community engagement and advocacy is crucial to developing an industry whose impacts are likely to be felt before its social benefits materialise. Crucially, local communities could be supported by Social Sciences and Humanities (SSH) research in creating new mythologies and imaginaries through which MRE technologies become an integral part of their culture, as well as part of their biophysical environment. A complex physical, political, and legal environment provides the context for these new marine energy technologies, and its development provides opportunities for SSH research to address issues around the sea and to integrate into the design of new marine energy seascapes.

Keywords Marine energy • Engagement • Mythologies • Disparities • Communities • Tidal • Wave

R. Brennan
Trinity Centre for Environmental Humanities, School of Histories and Humanities, Trinity College, Dublin, Ireland
e-mail: Ruth.Brennan@tcd.ie

R. Howell
School of Social and Political Science, Edinburgh University, Edinburgh, UK
e-mail: rhys.howell@ed.ac.uk

M. Graziano
Department of Geography and Environmental Studies, Central Michigan University, Mount Pleasant, MI, USA
e-mail: grazi1m@cmich.edu

A. M. O'Hagan
MaREI Centre, Environmental Research Institute, University College Cork, Cork, Ireland
e-mail: a.ohagan@ucc.ie

D. van der Horst
Institute of Geography and the Lived Environment, Edinburgh University, Edinburgh, UK
e-mail: Dan.vanderHorst@ed.ac.uk

G. Wright
Institute for Sustainable Development and International Relations, Paris, France
e-mail: glen.wright@iddri.org

B. Wynne
Department of Sociology, Lancaster University, Lancaster, UK
e-mail: b.wynne@lancaster.ac.uk

3.1 Introduction

Marine Renewable Energy, generating electricity from the movement of either the waves or tides, is a developing industry, with offshore commercial deployment of small arrays of devices now underway. For example, the MeyGen project in the north of Scotland has 6 MW on-grid capacity in the water, generated by four tidal energy devices, and is planned to expand to 398 MW. Around the world, there are around 40 open-sea test facilities for MRE. For the industry in the North Atlantic and elsewhere to grow, we must understand its relationship with our coastal communities. While academic research on land-based renewables abounds, the turn towards the sea is still in its infancy.

EU energy policy has been highly successful in making 'first generation' renewable technologies (e.g. solar and wind) commercial. Maritime policy is now focused on the innovation and Blue Growth of 'next generation' ocean energy. With an enabling regulatory framework, this technology could supply 10% of the EU's power demand by 2050 (Ocean Energy Europe (OEE) 2016). As a recent use of marine space, MRE raises not only scientific and technical challenges but also social challenges in places with deep physical, psychological, and spiritual connections to the sea. What are the effects of marine renewables on seascapes and the marine environment (Haggett 2008; Ladenburg 2008)?

Social Sciences and Humanities (SSH) research can improve the design/assessment of, and interaction with, complex sociotechnical issues, such as the energy transition, yet it remains underutilised in energy policy, especially in a marine context. To address this, members of the International Network for Social Studies of Marine Energy (ISSMER), an academic network formed to engage with this issue, held a two-day workshop in February 2018 in Edinburgh. Four representatives from MRE were invited, and together we discussed the response of the nascent MRE industry to social issues and considered the role and outlook for SSH research. Our four guest experts, representing MRE developers, government, and local community, engaged in lively and enlightening discussions with ISSMER researchers. This paper summarises that discussion.

Each guest posed a 'big question' to start the conversation, which was then directed towards key research domains. From these exchanges, five important themes emerged: *rights and ownership, community mythologies, disparities, design, and the need for an ecology of approaches*. These themes, presented below, reflect a broad range of important SSH factors relevant to MRE development. We include quotes from stakeholders; however, in accordance with the request of some stakeholders, their identity is kept confidential. In our conclusion, we reflect on the need for *sustained engagement* and the potential for advocacy by SSH.

3.2 Rights and Ownership

> "We need to create a sense, or reality, of ownership"—guest expert on the imperative for community control over resources.

MRE often requires exclusive use of marine space, since devices are anchored to the seabed and are, for practical purposes, permanent. Rights and ownership issues take diverse forms in the marine environment, from legal ownership of the devices themselves to the *de facto* ownership of the sea and seabed (backed by a lease or consent from the State). Certain common rights in the sea, such as fishing and navigation, may also have a legal underpinning. Many coastal communities place value on their traditional rights of access to maritime resources, which may not be codified in law, for example, First Nation rights (Wright et al. 2016). Physical and emotional proximity to the sea can generate a powerful sense of ownership. In many coastal societies, the sea is inextricably linked to community and identity. Some Pacific island communities see the environment, people, and custom as bound in a single concept, 'vanua', with no clear distinction between land and sea (Batibasqa et al. 1999). Communities may feel their common rights and well-established relationship with the sea are disavowed when a new industry disrupts their marine experience, by introducing visual or other changes, or by blocking access to their coastline. The MRE industry brings the tensions between these differing notions of ownership and relationality to the fore.

Whereas terrestrial planning systems have evolved around existing patterns of privately owned land, the situation at sea is more complex (Jay 2010). Landowners can generally use their land as they wish, with the government impinging on these private rights only where necessary to preserve legal order and protect the public interest (Johnson et al. 2013). At sea, States claim sovereign jurisdiction over their coastal waters, and private ownership of marine spaces remains rare. In many places, the sea is considered a commons, public good, or free for all to use, and legal regimes generally reflect this (Smith et al. 2012). The perception of the marine environment as being a 'public good' is even stronger in cultures or communities with close connections to the sea, precisely the communities that are seeing new industries develop on their shore. As the Blue Economy grows and seeks new capital opportunities, what was once considered a commons is being enclosed, as leases for maritime activity, such as MRE, are granted.

MRE developers occupy a peculiar position in a local community. They have quasi-ownership of a sea area (through their lease) and yet the seas are owned by the State on behalf of its citizens. MRE is generally located close to shore, with significant onshore building works and associated social and environmental impact. Resistance by a local community to MRE could close both business case and national resource. There is little margin for error in the development of MRE, especially tide energy, for which there are only a few viable locations.

Developers must acquire a lease to access the tide or wave resource, and also work with communities to ensure that their energy generation does not impinge upon local 'moral' ownership. Failure by a developer to take into account changing patterns of ownership, or initial suspicion by a local community towards MRE, can quickly lead to protest and conflict (De Groot et al. 2018). While marine planning policies are evolving, the focus has tended to be on established activities or the tension between development and conservation (Jones et al. 2013).

The question of ownership has implications beyond access to sea space. In some regions, direct financial payments to nearby communities have emerged as a way of easing tensions (e.g. wind power in Scotland). Alternatively, the State has granted rights over resources directly to communities, providing income in the form of rent or profits (Kerr et al. 2017). Communities must then discharge the decision-making responsibilities that come with such rights, leading to additional benefits in the form of increased social cohesion and empowerment (Rennie and Billing 2015).

A community may take figurative 'ownership' of a particular MRE device or project. Indeed, the origin stories of many renewable energy industries are strongly rooted in particular places and instil pride in their communities (Devine-Wright 2009). This personal connection might create a willingness to compromise in disputes over marine space.

While developers need only meet statutory requirements to acquire legal rights to occupy the sea space and exploit marine energy resources, they must also balance this with the rights and prerogatives of other sea users. There is little precedent, here, as few commercial developments have completed the planning process. It would be useful to develop some guidance. For example, maintaining a 'Social Licence to Operate' (SLO) could require community engagement and recognition of important values beyond the minimum regulatory requirements, and stretch inland, as infrastructural needs of the electricity grid emerge.

In short, there is growing public concern and research around ownership of maritime resources (Kerr et al. 2015). 'Blue Growth' industries like MRE are transferring rights of access and ownership from commons to private ownership. Social Sciences have an important role to play in understanding the tension between the legal rights of individual developers and strongly held 'sense' of ownership experienced by many coastal communities.

3.3 Community Mythologies

"It's getting to the stage where it becomes a part of them"—guest expert reflection on how marine renewable development can fit with how local communities interpret and imagine the world.

Mythologies and imaginaries are ways of understanding the world that help us to make sense of complex social issues (Anderson 1999; Levy and Spicer 2013). They are "imaginative patterns, networks of powerful symbols that suggest particular ways of interpreting the world" (Midgley 2004, p. 1). Existing mythologies around the sea influence how MRE technologies are received by communities. For example, the story of a small group of blacksmiths and teachers in West Jutland who became the Danish wind energy industry, backed by the Danish Government, provides a powerful 'from the people' origin myth for Denmark's wind industry (Graziano et al. 2017). How can communities be supported in creating new imaginaries through which MRE technologies become an integral part of their cultural as well as biophysical environments?

Since MRE is still in development, its mythologies have not yet been defined. This presents an opportunity for coastal communities to shape MRE mythologies appropriate to their particular socio-cultural context. In contrast, mythologies embedded in unsustainable investor hype around the industry (necessary for attracting financial support) should not be misinterpreted by communities as a likely source of local jobs or income. SSH researchers can manage MRE mythologies-in-the-making: for example, the future freezing of renewable energy subsidies in the UK could affect the way communities and the public view these technologies.

Communities need to be supported by consistent, locally rooted, and enduring mythologies around new energy technologies, particularly where they may be perceived as creating benefits for some (such as device developers) and obstacles for others (such as fishers displaced from fishing grounds). Establishing a culture and practice of sharing stories about these new technologies between marine communities (regionally, nationally, and across Europe) could be fundamental to the creation of new futures that speak to people at a grassroots level. Pioneering communities who have hands-on experience of new energy technologies could convey their nuanced understanding to those communities following in their footsteps. Their voices are more likely to be seen as untainted by the profit motive of developers. Communities can also prepare their mythologies for what might happen in their sea, as the industry matures in the years to come. Thus, a mythology could embrace the potential for co-existence with developers, and community ownership of devices in the future.

We propose creating a new cohort of SSH-informed marine 'architects' (local community 'designers' as distinct from statutory planners), who could ensure that socio-cultural issues, including mythologies, are embedded in MRE practices and policy from the outset. This could help communities, policymakers, and developers to recognise and accommodate an ecology of different relationships in bringing this new industry to maturity.

3.4 DISPARITIES

"Tidal regions must be developed intelligently to make best use of the resource"—guest expert on the small number of locations with tidal energy resources and how we must develop each one with care.

MRE operates across a complex physical, political, and legal environment, and there are a number of disparities that affect project development. These form opportunities for SSH research to inform and guide the intersection between industry, government, community, and environment.

Coherent marine management and planning is a recent undertaking, as governments grapple with Blue Growth. Marine governance has been driven by the dual forces of economic development and conservation. MRE devices are being developed by for-profit firms, potentially both contributing to, and clashing with, conservation objectives.

An additional disparity arises between the treatment of fossil fuels and renewables. While it is widely acknowledged that rapid de-carbonisation of our energy systems is necessary, path dependencies and established subsidies for fossil energy often mean that renewables are at a disadvantage. MRE developers feel hamstrung by burdensome environmental monitoring requirements, implemented due to uncertainty regarding their environmental interaction. By contrast, oil and gas projects benefit from decades of often state-supported offshore fossil fuel extraction. Even offshore wind has fewer environmental designations and existing maritime activities to contend with, since it is situated further out to sea.

Whereas communities have been developing onshore wind projects for decades, there is less potential for the development of community-owned MRE due to its high risk. Marine energy is likely to remain central-government and/or private investment owned for the near future.

The highly localised character of MRE makes it difficult to draw lessons from one project, community, or country, which can benefit others. For example, the positive community narrative regarding marine energy in Orkney, Scotland, which has seen considerable investment in both projects and community engagement, is different from projects elsewhere (see comparison between Orkney and Denmark in Watts and Winthereik 2018). There are disparities between communities and environmental contexts where projects are proposed.

MRE technologies also face different challenges. Tidal stream technologies (generating electricity from tidal flow) are the most advanced, but the worldwide availability of exploitable resources is limited. Wave energy technologies are yet to coalesce around a particular design, but there are more potential sites of resource. MRE technologies suffer from disparities in their spatiality and timing, and expectations that hold for tide energy might not hold for wave energy.

Compounding these disparities is the resource expectation. Terrestrial sources of renewable energy, such as wind, are relatively abundant, allowing developers a certain level of flexibility in selecting appropriate sites. By contrast, the number of potential sites for marine energy is much more limited. This means that developers must develop projects in those specific places and ensure that the community supports their projects. Community engagement, and a well-established socio-cultural relationship with marine energy, is therefore vital. As such, SSH concerns are central to negotiating across the many disparities in MRE development.

3.5 Design

"We need new stories. It doesn't matter if they are 'right' or not. What matters is the creation of fresh stories and ideas"—guest expert on community consultation as part of the design process for shaping future energy infrastructure.

The design process for MRE is, at present, focused on environmental and technical concerns, but social and cultural issues also emerge at each step of the design, development, and deployment process. MRE devices require environmental, technical, and social issues all to be resolved, and these are both related together and relational by nature. How might we design MRE projects to ensure good relations between all these aspects? There are resources in science studies, and other fields that specialise in social and technical relations, to support such a design process for energy infrastructures (Gabrys 2014; Watts 2014; Watts et al. 2018).

Experts and engineers can often miss the needs of people and publics. Device design is often a technical and proprietary process, but we might open these design processes to allow input and ideas from local experts, repositioning local communities as experts in their own seascape and its relations. Developers often seek to manage expectations of existing sea users and local communities, so they might inquire more broadly as to whether the community is open to MRE projects, and what expertise is available to them. Unlike wind energy, MRE design and implementation can be contingent on the complex sea environment, with many unknowns and scientific uncertainties. Local mariners (fishers, aquaculture workers, divers) are often the keepers of this local knowledge. Design processes could integrate this knowledge early on, helping to de-risk the outcome. Inquiries into local expertise and reception could highlight whether consent is likely or if there will be substantial resistance due to local mythologies and imaginaries.

MRE developers could learn from well-established approaches such as participatory design, co-design, and speculative design, which democratise design by emphasising the importance of location and participation of users and communities (Ehn et al. 2014; Kimura and Kinchy 2016).

The current statutory consultation process has rigid and specific legal frameworks, which can result in short bursts of intense community engagement and 'tick-box consultation'. However, device developers, conscious of the importance of their long-term relationship with a local community, would prefer a more expansive consultation. This would potentially de-risk projects. As one participant put it at our workshop, developers should "arrive early, and stay late" in a local community.

A related issue is how consultation is often conducted by a wide variety of third parties (often diverse consultancies, working on behalf of many different organisations). While this adequately fulfills legal requirements, it can fail to provide valuable feedback. A single, local point of contact between MRE developers and communities, such as a liaison person or a local 'champion' embedded in the community, could help bridge this gap.

Connecting back to MRE mythologies, a participatory design process could encompass social and cultural heritage and histories. This would affect how a new device forms a relationship to a local community and becomes accepted or not. For example, a MRE device could connect to existing cultural heritage of the sea, prior histories of energy extraction on land, sea ownership and rights disputes, or even public stories about the organisations, investors, or developers in circulation around the world. Reimagining the design process would allow for positive engagement in difficult socio-cultural issues. SSH, from design and policy engagement to cultural research and arts projects, could become the vanguard for engaging with, and making visible, the existing cultural context for sea energy.

3.6 Ecology of Approaches

> "Decisions are ultimately qualitative ... Putting numbers on something is usually a justification for what is a socially-driven decision"—guest expert on the limits of quantitative decision-making.

Rather than relying on a narrow set of methods, we propose an 'ecology of approaches' with diverse forms of evidence to understand the social impacts and relations in MRE. We should bring together quantitative analysis and qualitative methods, such as participatory mapping, ethnography, and cultural histories, to speak across industry, policy, and communities.

Different forms of narration, evidence, and language are used by different stakeholders for communications and knowledge transfer. SSH researchers have an important role to play in translating between these languages, bringing greater clarity to the views of stakeholders and conveying respect for their multiple knowledge systems. We propose going beyond merely quantifying environmental impacts towards understanding and translating the rich social and environmental interactions with MRE technologies (Harvey et al. 2016).

MRE is a start-up industry, with financial investment often coming from venture capital. This carries the risks of hype and disappointment, as well as miscommunication. The sociological literature on expectation and anticipation demonstrates that these stories and predictions about future industry are performative (Brown and Michael 2003). Narratives told by industry to local communities, investors, and policymakers are crucial in changing the future of the industry. However, these three domains need different stories and evidence, since they have diverse concerns and objectives. SSH has a role to play here. SSH researchers have established methods for supporting collaboration between local communities and new industries, for example, semi-structured interviews, observational surveys, and ethnography. Secondly, SSH can document the diverse and rich maritime relations and history that all participants have, using a range of both qualitative and quantitative evidence. As deployment is always specific to a place, this evidence can be used to improve our understanding of how the industry can become sensitive to local histories, as well as how local communities can be sensitised to future industry. SSH also provides access to a global range of socio-cultural approaches taken to MRE to reduce perceived risk (Wright et al. 2018).

SSH, particularly Arts and Humanities, has approaches that can create sea and energy stories within a local community that can establish an overt relationship with MRE ('prime the area'), enabling a local community to be ready to engage with marine energy deployment. For two examples, see participatory story-mapping (Brennan 2018), a method which could be used to improve developers' knowledge of local resources, and an energy walk used to develop local public engagement with sea energy (Winthereik et al. in press).

Overall, the variety of different relationships with the sea and energy necessitate the use of equally varied approaches to research. We need to create bespoke methods with different options and timeframes, aiming to move beyond an 'us versus them' mindset, to an MRE 'development community' that includes developers, researchers, policymakers, and the local community.

3.7 Conclusion

"Show you've learnt, and hang around late in conversation so you can show that you've learnt"—guest expert on previous effective community engagement.

The ecology of approaches to 'ownership' and the 'mythologies' created by communities around MRE should not be thought of as self-contained events in time. Each failed or successful project leaves a legacy that contributes to future discourses. As MRE technologies, particularly tidal, are bound to a handful of specific locations, these legacies are of great importance. The development of an effective institutional memory can help avoid issues resulting from over-information and under-preparation of the actors involved (Alavi and Leidner 2001). This 'memory' is more than a simple collection of papers on previous applications, which are already somewhat available. Along with a strengthening of the record-keeping process, we propose that such a collective memory could be generated and maintained through an institution devoted to preserving and circulating information regarding the experiences of MRE developers (Corbett et al. 2017).

The future has an important role for marine energy, especially in a context of Blue Growth. As one of the workshop stakeholders said, commenting on the need to sensitise people with evidence-based information: "We have time to do this as this industry is not ready now". Communities in energy-rich waters have been exposed to partial information as developers have attempted to establish their presence in recent years. These experiences may generate positive and negative expectations of future development based on partial information flows generated outside of the public sphere.

Within a context of Blue Growth Strategy, anticipating and sustaining a dialogue with the relevant communities will create the social capital necessary to justly support the diffusion of marine energy. But how do we connect the past with this future?

Mythologies created by communities are one important catalyst. The interconnected work of geographers, sociologists, economists, and anthropologists has demonstrated how social interactions, mediated by institutions, can support the diffusion and development of new innovations (Brown 1981), including renewable energy technologies (Firestone et al. 2009; Graziano and Gillingham 2015). SSH can help developers and scientific institutions to develop that 'reflexive discourse' (Wynne 1992; Wynne 2006) necessary for creating long-lasting trust between all parties involved during the emergence of MREs. As mythologies change, and as memories of successful and failed projects accumulate, social scientists can fill the void between appearing and disappearing stakeholders.

One 'tool' for filling such a void could be found in the concept of 'bridger organisations' (Wilson and MacDonald 2018). These communicate across organisational, sectoral, and national boundaries to preserve memories, and transmit knowledge over extended time. The form of bridgers can vary depending on the underlying policy landscape, from non-governmental organisation to research institutions to independent state-run agencies.

A bridging institution of this kind for marine energy will play two roles. First, it can collect and synthesise the forms and materials of past development processes. Alone, this role is insufficient to guarantee that knowledge is passed on. Therefore, secondly, this institution can participate in sustained advocacy and engagement with the local communities, preparing and keeping them informed, and collecting and managing their changing expectations, concerns, and requests. Further, this institution could assist developers in their relationship-building process, lowering the risks associated with developing concepts of ownership, recording mythologies, and formulating a seascape where the spatial and temporal disparities can be understood by all stakeholders, thus operationalising the knowledge it has preserved.

On the basis of this, we have outlined some key opportunities for SSH in supporting MRE (Box 3.1).

Box 3.1 Key opportunities for SSH-supported Marine Renewable Energy (MRE)

- Create a bridger organisation for MRE to support enduring international knowledge.
- Understand different notions of ownership that underpin potential marine resource conflicts.
- Facilitate respectful collaborations across different knowledge systems and forms of evidence, and develop the institutions and processes that could bring all actors together to create shared visions for the local deployment of these nascent technologies.
- Develop and apply contextual information to MRE development, rather than extrapolating the findings from one technology or location to another.
- Take inclusive and creative approaches to design, accounting for different interests, knowledge systems, and geographies. Such design processes could help in developing a constructive, ongoing narrative of shared values and benefits and the co-ownership of sociotechnical innovations.

Overall, technocratic strategies for the terrestrial deployment of renewable energy have often met with significant opposition and delay due to a lack of meaningful engagement with different community and stakeholder groups. SSH and socio-cultural approaches can inform and intercede in Blue Growth, to limit the risk of similar problems occurring in marine energy development and to make both sustainable communities and sustainable energy.

REFERENCES

Alavi, M., & Leidner, D. (2001). Review: Knowledge Management and Knowledge Management Systems: Conceptual Foundations and Research Issues. *MIS Quarterly, 25*(1), 107–136.
Anderson, B. (1999). *Imagined Communities* (2nd ed.). London: Verso.
Batibasqa, K., Overton, J., & Horsley, P. (1999). Vanu: Land People and Culture of Fiji. In J. Overton & R. Scheyvens (Eds.), *Strategies for Sustainable Development Experiences from the Pacific*. London: Zed Book.
Brennan, R. (2018). Re-storying Marine Conservation: Integrating Art and Science to Explore and Articulate Ideas, Visions and Expressions of Marine Space. *Ocean & Coastal Management*. https://doi.org/10.1016/j.ocecoaman.2018.01.036
Brown, L. (1981). *Innovation Diffusion*. New York: Methuen.
Brown, N., & Michael, M. (2003). A Sociology of Expectations: Retrospecting Prospects and Prospecting Retrospects. *Technology Analysis & Strategic Management, 15*(1), 3–18.
Corbett, J., Grube, D. C., Lovell, H., & Scott, R. (2017). Singular Memory or Institutional Memories? Toward a Dynamic Approach. *Governance, in press*. https://doi.org/10.1111/gove.12340.
De Groot, J., Campbell, M., Reilly, K., Colton, J., & Conway, F. (2018). A Sea of Troubles? Evaluating User Conflicts in the Development of Ocean Energy. In G. Wright, S. Kerr, & K. Johnson (Eds.), *Ocean Energy Governance Challenges for Wave and Tidal Stream Technologies* (pp. 170–190). London: Earthscan.
Devine-Wright, P. (2009). Rethinking NIMBYism: The Role of Attachment and Place Identity in Explaining Place-protective Action. *Journal of Community and Applied Psychology, 19*, 426–441.
Ehn, P., Nilsson, E., & Topgaard, R. (2014). *Making Futures: Marginal Notes on Innovation, Design, and Democracy*. Cambridge, MA: MIT Press.
Firestone, J., Kempton, W., & Krueger, A. (2009). Public Acceptance of Offshore Wind Power Projects in the USA. *Wind Energy, 12*(2), 183–202.
Gabrys, J. (2014). A Cosmopolitics of Energy: Diverging Materialities and Hesitating Practices. *Environment and Planning A, 46*(9), 2095–2109.

Graziano, M., & Gillingham, K. (2015). Spatial Patterns of Solar Photovoltaic System Adoption: The Influence of Neighbors and the Built Environment. *Journal of Economic Geography, 15*, 815–839.

Graziano, M., Musso, M., & Lecca, P. (2017). Historic Paths and Future Expectations: The Macroeconomic Impacts of the Offshore Wind Technologies in the UK. *Energy Policy, 108*, 715–730.

Haggett, C. (2008). Over the Sea and Far Away? A Consideration of the Planning, Politics and Public Perception of Offshore Wind Farms. *Journal of Environmental Policy and Planning, 10*(3), 289–306.

Harvey, P., Bruun Jensen, C., & Morita, A. (Eds.). (2016). *Infrastructures and Social Complexity: A Companion*. New York: Routledge.

Jay, S. (2010). Built at Sea. *Town Planning Review, 81*(2), 173–191.

Johnson, K., Kerr, S., & Side, J. (2013). Marine Renewables and Coastal Communities—Experiences from the Offshore Oil Industry in the 1970s and Their Relevance to Marine Renewables in the 2010s. *Marine Policy, 38*, 491–499.

Jones, P., Qiu, W., & De Santo, E. (2013). Governing Marine Protected Areas: Social–ecological Resilience Through Institutional Diversity. *Marine Policy, 41*, 5–13.

Kerr, S., Colton, J., Johnson, K., & Wright, G. (2015). Rights and Ownership in Sea Country: Implications of Marine Renewable Energy for Indigenous and Local Communities. *Marine Policy, 52*, 108–115.

Kerr, S., Johnson, K., & Weir, S. (2017). Understanding Community Benefit Payments from Renewable Energy Development. *Energy Policy, 105*, 202–211.

Kimura, A. H., & Kinchy, A. (2016). Citizen Science: Probing the Virtues and Contexts of Participatory Research. *Engaging Science, Technology, and Society, 2*(0), 331–361.

Ladenburg, J. (2008). Attitudes Towards on-land and Offshore Wind Power Development in Denmark; Choice of Development Strategy. *Renewable Energy, 33*(1), 111–118.

Levy, D., & Spicer, A. (2013). Contested Imaginaries and the Cultural Political Economy of Climate Change. *Organization, 20*(5), 659–678.

Midgley, M. (2004). *The Myths We Live By*. London: Routledge.

Ocean Energy Europe (OEE). (2016). *Ocean Energy Strategic Roadmap: Building Ocean Energy for Europe*. Brussels: European Commission.

Rennie, F., & Billing, S. (2015). Changing Community Perceptions of Sustainable Rural Development in Scotland. *Journal of Rural Community Development, 10*(2), 35–46.

Smith, H. D., Ballinger, R. C., & Stojanovic, T. A. (2012). The Spatial Development Basis of Marine Spatial Planning in the United Kingdom. *Journal of Environmental Policy and Planning, 14*, 29–47.

Watts, L. (2014). Liminal Futures: A Poem for Islands at the Edge. In J. Leach & L. Wilson (Eds.), *Subversion, Conversion, Development: Cross-Cultural Knowledge Exchange and the Politics of Design*. Cambridge MA: MIT Press.

Watts, L., & Winthereik, B. R. (2018). Ocean Energy at the Edge. In G. Wright, S. Kerr, & K. Johnson (Eds.), *Ocean Energy: Governance Challenges for Wave and Tidal Stream Technologies*. Abingdon and New York: Routledge.

Watts, L., Auger, J., & Hanna, J. (2018). The Newton Machine: Reconstrained Design for Energy Infrastructure. In P. Sumpf & C. Büscher (Eds.), *Control, Change and Capacity-Building in Energy Systems: SHAPE ENERGY Research Design Challenge* (pp. 135–142). Cambridge: SHAPE ENERGY.

Wilson, L., & MacDonald, B. H. (2018). Characterizing Bridger Organizations and Their Roles in a Coastal Resource Management Network'. *Ocean and Coastal Management, 153*, 59–69.

Winthereik, B. R., Maguire, J., & Watts, L. (in press). The Energy Walk: Infrastructuring the Imagination. In J. Vertesi & D. Ribes (Eds.), *Handbook of Digital STS*. Princeton, NJ: Princeton University Press.

Wright, G., O'Hagan, A.M., de Groot, J., Leroy, Y., Soininen, N., Salcido, R., Castelos, M., Jude, S., Rochette, J., & Kerr, S. (2016). Establishing a Legal Research Agenda for Ocean Energy. *Marine Policy, 63*, 126–134.

Wright, G., Kerr, S., & Johnson, K. (Eds.). (2018). *Ocean Energy: Governance Challenges for Wave and Tidal Stream Technologies*. Abingdon and New York: Routledge.

Wynne, B. (1992). Misunderstood Misunderstanding: Social Identities and Public Uptake of Science. *Public Understanding of Science, 1*(3), 281–304.

Wynne, B. (2006). Public Engagement as a Means of Restoring Public Trust in Science—Hitting the Notes, But Missing the Music? *Public Health Genomics, 9*(3), 211–220.

Open Access This chapter is licensed under the terms of the Creative Commons Attribution 4.0 International License (http://creativecommons.org/licenses/by/4.0/), which permits use, sharing, adaptation, distribution and reproduction in any medium or format, as long as you give appropriate credit to the original author(s) and the source, provide a link to the Creative Commons license and indicate if changes were made.

The images or other third party material in this chapter are included in the chapter's Creative Commons license, unless indicated otherwise in a credit line to the material. If material is not included in the chapter's Creative Commons license and your intended use is not permitted by statutory regulation or exceeds the permitted use, you will need to obtain permission directly from the copyright holder.

CHAPTER 4

Looking for Perspectives! EU Energy Policy in Context

Anna Åberg, Johanna Höffken, and Susanna Lidström

Abstract Transitioning to less carbon-intensive energy systems involves making difficult choices and priorities. This chapter imagines three individuals who are affected in different ways by EU energy policy. Their fictional stories illustrate that energy policies are embedded in social, historical and cultural practices and need to take a broader perspective than either technological fixes or a narrowly defined goal of low or zero carbon emissions to be fair and effective. We argue that this is often not reflected in the EU's energy policy frameworks, and use the Energy Roadmap 2050 to demonstrate our point. Contrary to the impression

---

A. Åberg (✉)
The Division for Science, Technology and Society, Chalmers University of Technology, Göteborg, Sweden
e-mail: anna.aberg@chalmers.se

J. Höffken
School of Innovation Sciences, Eindhoven University of Technology, Eindhoven, Netherlands
e-mail: J.I.Hoffken@tue.nl

S. Lidström
Division of History of Science, Technology and Environment, KTH Royal Institute of Technology, Stockholm, Sweden
e-mail: suslid@kth.se

© The Author(s) 2018
C. Foulds, R. Robison (eds.), *Advancing Energy Policy*,
https://doi.org/10.1007/978-3-319-99097-2_4

given by the roadmap, a narrow technocratic empirical basis for a policy is not enough to define and solve an energy problem. Energy issues are societal problems and need to be addressed as such.

Keywords Energy poverty • Energy production • Climate change • Market liberalisation • Renewable energy

4.1 Introduction

For the European Union to transition to a less carbon-intensive energy system, difficult choices need to be made about different renewable energy sources and their effects on regions, nations and citizens of the EU and beyond. This chapter examines this simple yet complex point and argues that energy policy frameworks tend to disproportionally focus on technological aspects of possible energy futures while paying less attention to the social embeddedness of energy production and consumption. We demonstrate our argument through a close reading of the EU's Energy Roadmap 2050 (European Commission 2012). To ground our analysis, we begin by imagining three individuals who are affected in different ways by EU energy policy. Their stories, though fictional, are grounded in actual events and supported by relevant literature (as per this chapter's three endnotes). From these stories, we proceed to reflect on how the issues they attend to are—or are not—accounted for in EU energy policy frameworks.

4.2 Alva, Daniela and Ambika

The setting for our fictional stories is a citizen platform organised in collaboration with the European Commission a few years into the future, where three women have been invited to give their perspective on the EU's Energy Roadmap 2050.

"Welcome everyone! We, the organisers of this citizen-platform, are happy to see that so many of you have come to join us. This year, 2021, marks the tenth year after the launch of the Energy Roadmap 2050. Reason enough for the Commission to take stock and to review the ambitions set out in the roadmap. Importantly, the insights generated during this citizen-platform will inform the Commission's review process." (Box 4.1)

> **Box 4.1 Excerpt from the Citizens' summary of the Energy Roadmap 2050**
>
> - All citizens will benefit from lower greenhouse gas emissions, more secure and affordable energy if strategic decisions and investments are taken now to save energy, invest in low carbon energy sources and build intelligent and diversified energy networks.
> - The development of new energy alternatives will sustain Europe's competitiveness in growth and job-creating new industries.
> - Transforming the energy system will: empower consumers and make the energy bill more controllable and predictable; it will lead to more investment in the EU and lower bills for external fossil fuels; and it will increase energy security by more domestic supply.

'Even though the roadmap includes a summary for "citizens", we as citizens rarely get the opportunity to discuss in person in what ways these noble goals impact our local and national realities in practice.

The citizen platform has been organised to start such a discussion. We hope that it allows for perspectives to be heard that go deeper beneath the surface of the simplistic goals of the roadmap. For example: Will all citizens in fact benefit from the goals set out, or may some gain advantages at the expense of others? How do we make decisions when we need to choose between what is cheap and what is sustainable in our everyday lives? Where in the EU does growth take place? And in what ways is growth sustainable?

In order to explore at least some of these questions, we have invited three women to give their views on the roadmap. They represent groups whose voices are all too often peripheral in discussions on energy—but whose lives are centrally impacted by the decisions evolving from these discussions in which they did not take part.

So, let me introduce the three panellists sitting here next to me on the stage. They are Alva, from Sweden, Daniela from Bulgaria and Ambika from India.

Alva is a member of the Sami community, the indigenous people in northern Europe. The relationship between the Swedish government and the Sami people has a conflicted history, especially with regard to energy extraction and use, and Alva will address how energy policies adopted on the national level in Sweden may affect her community. Also on the panel is Daniela from Bulgaria. For her, energy security is perhaps a more important topic than sustainable energy, at least in the short term. This might also be true for our third invitee on the panel, Ambika, who has been invited to recognise that the effects of EU energy policy are inextricably linked to what happens outside Europe. She will reflect on this with a view on and from India.

I would like to open the discussion by asking you the panellists in what ways your realities mirror the goals set, almost ten years after the Energy Roadmap 2050 was laid out?'

A silence follows as Alva adjusts her notes and switches on the microphone speaker standing on the table in front of her. Then she starts speaking:

'*I live in northern Sweden, in the Swedish part of Sápmi, the traditional region of the Sami people. I am also a member of the Swedish Sami Parliament. A basic premise for me and for the Sami people is that for us, all questions are environmental questions. The natural environment is an integrated part of all aspects of our lives. There is no separation between our nature and our culture. This means that any destruction of the natural environment in Sápmi is a destruction of Sami culture as well.*

The Sami community is very concerned about climate change. As inhabitants of the north, we are likely to experience serious harm to our environment as temperatures rise. Like many other indigenous peoples around the world, we are exposed to changes in the climate because our lives and livelihoods are so closely tied to the landscapes around us. The incremental damage that long-term rising temperatures would inflict on the Sápmi region would be detrimental to Sami culture. If the natural conditions in our home change too much, it will be impossible for many of us to maintain our way of life.

However, I am even more concerned about short-term exploitation and destruction of Sápmi. There is a long history of colonisation of Sami territories and resources by the Swedish state. The Swedish government refuses to recognise this, despite remarks from the United Nations. The colonisation is on-going and risks intensifying in the name of transitioning to renewable energy. This must be stopped.

An example is the case of Stekenjokk, located within Swedish Sápmi, close to the Norwegian border. Stekenjokk is a spiritually and culturally important place to the Sami. We use it for traditional reindeer herding, and the local Sami community have constitutional rights to use the area for this purpose. Despite this, the Swedish state has moved forward with plans to allow private companies to develop large wind power plants in Stekenjokk. The

Sami people have neither been consulted nor informed about these plans, which amount to nothing less than an attempt at land-grabbing. Large-scale development of wind power would significantly affect the traditional use of the area by the Sami. A wind power plant causes a major disruption in the landscape. The area of the plant itself would no longer be suitable for reindeer herding. In addition, there would be additional changes in the form of new roads and other infrastructure and an increased number of people moving around in the area. In effect, Stekenjokk would become unrecognisable to the Sami.

There are other similar examples. They show that the second point in the "Citizens' summary" of the roadmap—that development of renewable energy will result in job creation and growth—is not true at all. At least not for all people. It is a very simplistic and idealistic statement, in my view. In Stekenjokk and other areas of Sápmi identified as suitable for wind power by the Swedish government, it would have the opposite effects—Sami jobs would be rendered impossible, and our economies would die out, not grow. This goes against principles set out in other documents and agreed to by the EU, which emphasise that development needs to be sustainable not only environmentally but socially as well; no one should be left behind.

I think that renewable energy sources are too often idealised and their contexts not sufficiently addressed. The roadmap should recognise that there are existing and potential conflicts around these types of developments and protect groups that are at risk of suffering from unjust and inappropriate locations of new energy plants. It is the view of the Sami Parliament, for example, that energy production should be primarily local and that therefore energy production sites such as wind power plants should be located first and foremost in southern Sweden, where most energy is consumed, instead of outsourced to areas in the north that may seem unoccupied but are in fact home to Sami people and important to their cultural and economic activities. When the Swedish government targets and exploits people within its own borders in this way, the EU should help to protect the Sami and their rights'.[1]

Alva pauses for some seconds and looks at Daniela, who starts speaking after switching on her microphone:

'Thanks, Alva. My story takes us from the EU's far north to the EU's far east: to Bulgaria.

I would like to start by reflecting on the consumer-empowerment aspect. According to the energy roadmap, transforming the energy system will empower consumers and make energy bills more controllable and predictable for the European consumer, like me. This is a worthy goal, but how will it be *achieved in practice? And what does empowerment mean, when you do not have any good choices to make? As an example, in Bulgaria, a big part of the housing is not sufficiently insulated, and the energy intensity is the highest in all of the EU. This means that a large amount of energy is needed to heat these houses, leading to high costs for us. My family used to have central heating, but the system is so expensive and dysfunctional, so we decided to leave the system and get electric heating instead. It is cheaper, and we can control our own heat and turn it down when we cannot afford more. Our choice, then, is between financially controllable insufficient heating and expensive non-functional heating. The roadmap tells me that individual European households will need to make investments in their housing to heighten the energy efficiency, but where will we get the money to do that? We can get financial help to pay our energy bills during the winter, but not enough for repairs of our houses.*

A couple of years ago, 64 per cent of the Bulgarian population reported that they were not able to keep their home sufficiently heated during winter, and 32 per cent that they had debts on their energy bills. According to the World Bank, 61 per cent of the population live in what they count as energy poverty. However, some say that this is due to the fact that the parameters are so different that they cannot be compared to other countries. This, of course, begs the question, if we are not like the other countries, why should we adhere to the same rules?

In the end, a big part of the population has gone back to firewood to be able to afford basic heating. Ironically, this is one of the reasons that Bulgaria reaches its Europe 2020 objectives for renewable energy, since firewood is counted as biomass, as opposed to fossil-fuel energy. Energy efficiency is quoted as one basic strategy to reach the EU energy goals. However, market liberalisation and more renewable energy have not helped us become more energy efficient. We need higher income levels and better infrastructure. Although housing quality has improved during the last couple of years, so far market liberalisation has only served to increase our energy bill, partly because of renewable energy which is more expensive. I have also heard that there has been a lot of corruption regarding the subsidies given to renewable energy projects, so I agree with Alva that there is an idealised view of renewables that does not fit my reality. And, speaking of market liberalisation, despite our so-called free market, I can only choose from one electric company anyway, since the old companies have a quasi-monopoly in the different regions of Bulgaria.

Part of the bigger problems that need to be overcome in Bulgaria in order to fight energy poverty are corruption and income levels. I do not see any solution to these problems in the EU roadmap. Instead, most of both the government policy and the EU-mandated energy market changes have meant a higher energy bill for my household, without addressing the underlying problems of Bulgaria's energy sector'.[2]

Ambika nods at Daniela and then faces the audience. After switching on her microphone, she says:

'Thanks for inviting me to this panel and thank you for sharing your views, Daniela and Alva.

In India, where I come from, the discourse on energy is almost always related to the imperative of the nation's economic growth. India needs economic development, considering that it is the country with the largest number of people living below the international poverty line. At the same time, India aims to become a global market player. Economic analysts, policymakers and business leaders dream about reaching double-digit growth rates and establishing India as the fastest growing major economy in the world market.

Indeed, it is a high-carbon development. Coal is clearly the main source of powering India's economy. And emissions keep increasing. But what is the alternative? Stop growing and stop lifting millions out of poverty? And just as an important side remark: A look back into history shows that India bears little responsibility for all the emissions that have built up over time and that went along with the carbon-intensive development of the Western world.

I know that the way I put it is a bit simplistic. If we consider current emissions, India is the third largest emitter and plays a crucial role in combatting climate change. There is indeed a sense of urgency: India is considered as one of the countries that will be most severely affected. Over time India has increasingly committed herself to taking an active stance in combatting climate change. But this, I feel, has not been met with similar engagement from developed countries. Sure, the EU and India have set up different programmes and issued joint declarations to combat climate change and engage in collaborative action plans. But are these engagements set up on equal grounds?

For example, the transfer of technology and funding from developed to less developed nations is essential for both mitigation and adaption to climate change. Just consider: India's population is projected to grow to 1.7 billion people in 2055. It really matters how India aims to power the life of her people. The transfer of technology, intellectual property rights and funding is needed to do this in a clean way.

But the EU has been reluctant to facilitate this transfer. Despite declarations, action plans and agreed obligations, the technology transfer is generally not offered at affordable prices, and developed countries do not provide the money they promised to finance climate change measures. I feel that these initiatives are just a way for the EU to open up a profitable outlet market for the EU's "green" products. I can also put it more bluntly: Is Europe again engaging in a form of neo-colonialism, where leaders pride themselves with "green" growth while keeping others dependent?

These are the thoughts which come to my mind when I read the second bullet point of the roadmap about Europe's aim to develop energy alternatives and stay competitive. I understand that EU policies are tailored towards issues directly relevant to the soon 27 member states. But the exclusive inward focus, or even selfishness, is irresponsible: historically, socially and ethically. And not least in the context of a changing climate—which will eventually affect us all, regardless where on earth we live'.[3]

4.3 Reflection

It is time to leave the citizen platform and turn towards the question of what can be learnt from the accounts of Alva, Daniela and Ambika.

To start with, their stories illustrate that whether a particular energy source is sustainable is context dependent. What is sustainable in one sense and for a certain group may not be so for another community, or on a different time horizon. While we do not argue against the need to transition to less carbon-intensive energy sources in order to mitigate climate changes, we do argue that complexities and perspectives surrounding the sustainability and desirability of different renewable energy sources are sometimes not sufficiently recognised in policy documents, for example, in the EU Energy Roadmap 2050. The roadmap paints a simplified picture of problems that are solvable through technical innovation and economic regulation 'for the benefit of all' ((European Commission 2012), p. 19). Our stories contradict this sweeping statement by bringing attention to the societal embeddedness of energy production and consumption 'on the ground' (also see Kerr et al., Chap. 3 in this book). Alva's account, for example, shows that plans for renewable energy tie in to histories of power relations and earlier resource exploitation in Sami territories. For her, development of wind power is a continuation of the internal colonisation of Sami lands practised by the Swedish state for centuries. In Ambika's account, too, histories of power relations emerge. Her story draws attention to how the relationship between the EU and India in contemporary climate agreements is coloured by questions around historical responsibilities and current possibilities for climate action and development.

In order to understand and approach these problems, a historical point of view is necessary, which includes a discussion about previous path dependencies and the consequences of colonial organisation. These issues have been tackled extensively within Science and Technology Studies (STS) and History of Technology. As an example, our aim with showing that renewable energy can have negative sides in some contexts is not to argue against renewables but to point out that mistakes made in the past need to be avoided in the transition to renewable energy sources, in order for them to be both socially and environmentally sustainable over a long period of time. The way that renewable energy is described in the roadmap—as more or less the solution to everything—is sometimes reminiscent of what within STS and history of technology is referred to as the idea of the 'technological fix', which is when narrow technological solutions are prioritised and applied even though the problem often lies in a political, economic, social and cultural system, often leading to new problems and non-efficient use of the technology (Bijker et al. 1987). This is reflected in the roadmap's focus on increased energy efficiency and other technological developments along with improved access for those currently considered energy 'poor', while very little is said about the possibility of decreasing the energy consumption of groups with very high consumption rates.

Within the EU, energy poverty has become a central issue for policy, and it is now mandatory for member countries to monitor energy poverty and report to the commission. However, there has not until recently existed any common EU practice to fight it (Middlemiss et al. 2018—Chap. 2 in this book). There are different ways to look at energy poverty, and poverty more broadly. One side is income rate in relation to energy prices, which is the World Bank view. However, there is also the issue of access to energy and to which kinds of energy. This view is more related to prioritising infrastructure and market development (Kisyov 2014). We see from Daniela's account that new infrastructure is needed to achieve a transition to renewables and to heighten energy efficiency, but the question of responsibility is still largely unsolved, and the state has to prioritise in regard to where and how to build. In the end, in this example, a lot of costs tend to fall on the table of the individual. On the other hand, as Alva's story highlights, new infrastructure can also turn out to be problematic.

How do we prioritise between supporting one lifestyle or the other? This question is also urgent in India, where Ambika reminds us that development and access to energy are vital for the well-being of millions.

Prioritisations are also an issue on an individual level. According to the roadmap, energy efficiency and lower prices will go hand in hand with a more sustainable energy sector. This may be true in the long run, but currently this is not the case for many citizens of Europe. Every choice people like Daniela are forced to make in their everyday life regarding their consumption of goods and energy can be seen as an exercise in goal conflict on several levels. Should family economics be prioritised, when consumer prices of renewables are more expensive than alternatives, or the climate? Whose goals are more important? The EU-level goals for climate and development? National or local goals? Individual ones? Thus, the tension between energy consumption and climate mitigation can be followed from the institutional level of the EU all the way down to the lives of its citizens. Consumers may also have to juggle information which may be incorrect or contradictory to their experience. While the view of renewables among some consumer groups in Bulgaria is reflected in Daniela's account, in reality issues surrounding renewable energy are more complex than in her narrative. For Daniela, however, the choice may still boil down to choosing the cheapest energy. This is not necessarily a simple economic choice but one that is embedded in social practice connecting her to a network of other individuals and institutions. A broad social theory which captures the full complexity of contextual choice can help change social practices and priorities of both policymakers and consumers (Shove 2014).

By using the roadmap as an example, we want to show that the technology focus of this particular policy framework clashes with the historical and social contexts that it will be applied to, and this can hamper its intention and enactment. As stated in the document, the European Commission will discuss future energy policy 'with other EU institutions, Member States and stakeholders on the basis of this roadmap' ((European Commission 2012), p. 19). For those discussions to be as fair, inclusive and effective as possible, the roadmap needs to recognise the dilemmas and diverse priorities of different groups and thereby provide a baseline for ensuing negotiations. For this reason, it is important to pay attention to how the roadmap frames the challenge of transitioning to a decarbonised energy system. As literary studies and related fields have shown, framing narratives shape ideas and discussions, for example, by recognising or neglecting certain groups and issues (e.g. Lakoff 2010). Acknowledging the complexity of an issue by inviting more than narrow techno-economic perspectives is a necessary step to make informed and inclusive decisions on what to prioritise and why. Ideally, energy policy frameworks such as the roadmap could enable policymaking that is based on 'contextualised prioritising' by weighing other-than-market considerations into the mix of decision variables. This would not take away the fact that hard decisions need to be made but expand and explicate the basis on which they are grounded.

4.4 Conclusion

This chapter started out with the accounts of Alva, Daniela and Ambika. While this is not the place to give final answers to the questions the women raise in their statements, their stories illustrate the complexity of energy policy and how the issues that face policymakers are not necessarily those that face energy consumers. When a narrow technocratic perspective is applied as an encompassing framework, a big part of the issue becomes invisible. This is also why the Energy Roadmap 2050 does not help our narrators; it only addresses a small part of the problem. Humanities and social sciences can make the whole map of complexities that lies behind an 'energy issue' more visible. This may lead to a broadening of what an 'energy problem' is, to encompass all the different social, political and cultural concerns that are often at the core of seemingly technical energy issues. Through such a redefinition, new relations and routes to problem solving can be envisioned.

Notes

1. References for Alva's account include European Commission (2001), Lawrence (2014), The Sami Parliament (2009a, b) and United Nations (2015).
2. References for Daniela's account include Kulinska (2017), Kisyov (2014), Martino (2015) and Pavlov (2018).
3. References for Ambika's account include Carrington and Safi (2017), Mohan (2017), European Commission (2011) and World Economic Forum (n.d.).

References

Bijker, W., Hughes, T., & Pinch, T. (1987). *The Social Construction of Technological Systems: New Directions in the Sociology and History of Technology*. Cambridge, MA: MIT Press.

Carrington, D., & Safi, M. (2017). How India's Battle with Climate Change Could Determine All of Our Fates. *The Guardian*, [online]. Retrieved April 27, 2018, from https://www.theguardian.com/environment/2017/nov/06/how-indias-battle-with-climate-change-could-determine-all-of-our-fates

European Commission. (2001). *A Sustainable Europe for a Better World: A European Union Strategy for Sustainable Development*. Brussels: European Commission.

European Commission. (2011). *Citizen's Summary: Energy Roadmap 2050*. [online]. Retrieved June 7, 2018, from https://ec.europa.eu/energy/sites/ener/files/documents/citizens_summary.pdf

European Commission. (2012). *Energy Roadmap 2050*. Luxembourg: Publications Office of the European Union.

Kisyov, P. (2014). *Report on National Situation in the Field of Fuel Poverty*. Work Package 2 report in the REACH project. [online]. Retrieved June 8, 2018, from http://reach-energy.eu/wordpress/wp-content/uploads/2014/12/D2.2-EAP_EN.pdf

Kulinska, E. (2017). Defining Energy Poverty in Implementing Energy Efficiency Policy in Bulgaria. *Economic Alternatives, 4*, 671–684.

Lakoff, G. (2010). Why It Matters How We Frame the Environment. *Environmental Communication, 2*(1), 70–81.

Lawrence, R. (2014). Internal Colonisation and Indigenous Resource Sovereignty: Wind Power Developments on Traditional Saami Lands. *Environment and Planning D: Society and Space, 32*, 1036–1053.

Martino, F. (2015). Green Energy in Bulgaria: An Uneasy Success. Osservatorio balcani e caucaso transeuropa. [online]. Retrieved June 8, 2018, from https://www.balcanicaucaso.org/eng/Areas/Bulgaria/Green-energy-in-Bulgaria-an-uneasy-success-158848#

Mohan, A. (2017). *From Rio to Paris: India in Global Climate Politics*. ORF Occasional Paper 130. New Delhi: Observer Research Foundation.

Pavlov, S. (2018). *Bulgaria Among EU Countries with Highest Consumption of Electricity from Renewables*. Radio Bulgaria, 12 February. [online]. Retrieved June 8, 2018, from https://www.bnr.bg/en/post/100932162/bulgaria-among-eu-countries-with-highest-consumption-of-electricity-from-renewables

Shove, E. (2014). Putting Practice into Policy: Reconfiguring Questions of Consumption and Climate Change. *Contemporary Social Sciences, 9*(4), 415–429.

The Sami Parliament. (2009a). *Eallinbiras*. [online]. Retrieved June 4, 2018, from https://www.sametinget.se/9008
The Sami Parliament. (2009b). *Sametingets syn på vindkraft i Sápmi*. [online]. Retrieved June 4, 2018, from https://www.sametinget.se/vindkraftpolicy
United Nations. (2015). *Transforming Our World: The 2030 Agenda for Sustainable Development*. [online]. Retrieved June 8, 2018, from http://www.un.org/ga/search/view_doc.asp?symbol=A/RES/70/1&Lang=E
World Economic Forum. (n.d.). *Why India Is Most at Risk from Climate Change*. Retrieved April 27, 2018, from www.weforum.org/agenda/2018/03/india-most-vulnerable-country-to-climate-change

Open Access This chapter is licensed under the terms of the Creative Commons Attribution 4.0 International License (http://creativecommons.org/licenses/by/4.0/), which permits use, sharing, adaptation, distribution and reproduction in any medium or format, as long as you give appropriate credit to the original author(s) and the source, provide a link to the Creative Commons license and indicate if changes were made.

The images or other third party material in this chapter are included in the chapter's Creative Commons license, unless indicated otherwise in a credit line to the material. If material is not included in the chapter's Creative Commons license and your intended use is not permitted by statutory regulation or exceeds the permitted use, you will need to obtain permission directly from the copyright holder.

PART II

Social Sciences and Humanities in Interdisciplinary Endeavours

CHAPTER 5

Achieving Data Synergy: The Socio-Technical Process of Handling Data

Sarah Higginson, Marina Topouzi, Carlos Andrade-Cabrera, Ciara O'Dwyer, Sarah Darby, and Donal Finn

Abstract Good quality research depends on good quality data. In multidisciplinary projects with quantitative and qualitative data, it can be difficult to collect data and share it between partners with diverse backgrounds in a timely and useful way, limiting the ability of different disciplines to collaborate. This chapter will explore two examples of the impact of data collection and sharing on analysis in a recent Horizon 2020 project, RealValue. The

---

S. Higginson (✉) • M. Topouzi • S. Darby
Environmental Change Institute, School of Geography and the Environment, University of Oxford, Oxford, UK
e-mail: sarah.higginson@ouce.ox.ac.uk; marina.topouzi@ouce.ox.ac.uk; sarah.darby@eci.ox.ac.uk

C. Andrade-Cabrera • D. Finn
School of Mechanical and Materials Engineering, University College Dublin, Dublin, Ireland
e-mail: carlos.andradecabrera@ucdconnect.ie; donal.finn@ucd.ie

C. O'Dwyer
School of Electrical and Electronic Engineering, University College Dublin, Dublin, Ireland
e-mail: ciara.o-dwyer@ucdconnect.ie

© The Author(s) 2018
C. Foulds, R. Robison (eds.), *Advancing Energy Policy*,
https://doi.org/10.1007/978-3-319-99097-2_5

main insight is that it is not only projects but also the processes within them such as data collection, sharing and analysis that are socio-technical. We shall examine two examples within the project—validating the models and triangulating the qualitative data—to examine data synergy across four dimensions: *time* (synchronising activities), *people* (managing and coordinating actors), *technology* (in this case focusing mainly on connectivity) and *quality*. Recommendations include developing a data protocol for the energy demand community built on these four dimensions.

Keywords Data collection and sharing methods • Socio-technical • Multidisciplinary • Energy demand • Demand response • Smart grid

5.1 Introduction

A large number of field trials have attempted to understand energy use in buildings (e.g. Economidou et al. 2011; Jones et al. 2013; TSB 2014; Guerra-Santin et al. 2013; Gupta and Kapsali 2015). Nevertheless, the number of studies with complete monitoring equally capturing building data, technologies and people is limited, a fact recognised by the Buildings Performance Institute Europe (BPIE) as limiting the impact of this research on European policy (Economidou et al. 2011). Notwithstanding their size, samples and research scope, many studies experience similar pitfalls in their data collection processes. Despite recognition of the need to combine multiple methods to understand the multidimensional socio-technical issues (Topouzi et al. 2016) and ongoing recognition of the ontological and language challenges of multidisciplinary work (Mallaband et al. 2017; Robison and Foulds 2017; Sovacool et al. 2015), there is less focus on the challenge of data collection and the implementation of these methodologies.

This chapter reflects on the socio-technical nature of data collection and sharing in multi-partner multidisciplinary[1] projects: not just the fact that different types of data need to be collected and analysed but the expectations different disciplines have of data[2] and the different skills they bring to the analysis. Recognising this and planning accordingly increases the chances of high-quality, useful data being used in collaborative ways in complex consortia. We suggest four dimensions to achieving data synergy[3] in such contexts: synchronising data processes in time, coordinating the people involved both logistically and in terms of their skills and expectations, recognising the multiplicity of issues affecting both social and technical data collection and paying attention to data quality.

Although the chapter will use examples from RealValue[4] (see Fig. 5.1), the issues discussed are common to most multidisciplinary projects with multiple actors. The chapter will use two illustrative examples. The first examines attempts to validate bottom-up models of energy demand using

> RealValue was a 3 year demonstration project (2015-2018) exploring the potential of Demand Response (DR) through the installation of Smart Electric Thermal Storage (SETS) space- and water-heating systems in several hundred properties (domestic and non-domestic) across trial sites in Ireland, Germany and Latvia. Whereas previously, storage heating typically only charged up overnight, the aim was to demonstrate how smart electric storage and water heating might support the functioning of the grid through Demand Response (DR) if it was able to switch on or off at any time (provided customers' needs were being met) in order to match demand with available supply.
>
> The project involved a multidisciplinary group of energy modellers, social scientists, manufacturers, engineers, software designers, network operators and the electricity supply industry and was divided into two strands: on-the-ground implementation, which collected data in properties, and a modelling component based on archetypal data and validated by trial data. Both strands started in parallel straight away. These fitted together as outlined in the diagram below, which shows the interrelationship between the two strands (to be achieved through data sharing); the importance of timing (given the need to synchronise the strands to produce deliverables within the project time-frame); and the difference between the original plan of the project and what actually happened (which is discussed in more detail later). This is a fairly standard project framework but has inherent difficulties built into the data collection process, which is what this paper addresses.

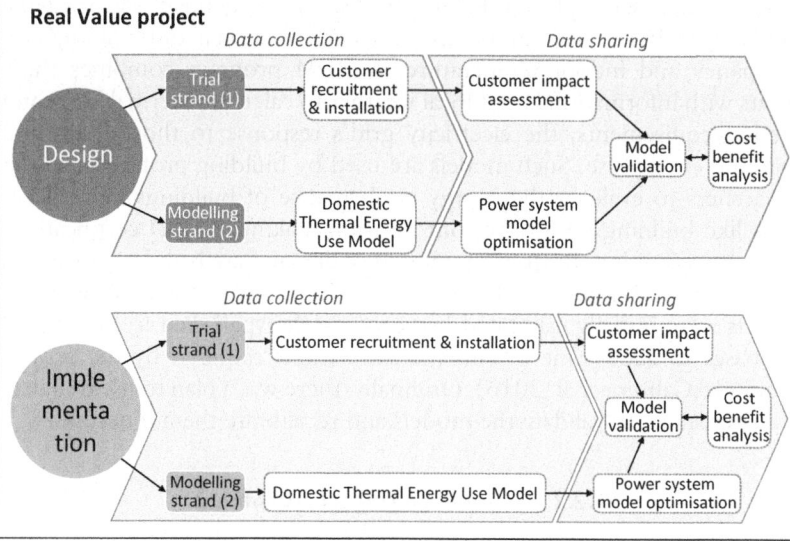

Fig. 5.1 Project description, design (top flowchart) and implementation (bottom flowchart)

trial data collected during the project. The second explores efforts to triangulate the qualitative data collected on customers, using monitoring data from the heating and hot water appliances fitted in their homes.

The chapter will start by introducing the background context of the project, move on to discussing the four dimensions of data synergy and finish with some recommendations for achieving data synergy.

5.2 Background Context

In order to later appreciate the data requirements of each project strand, it is necessary to describe them briefly.

5.2.1 Modelling

The plan for the modelling work was to integrate a building energy model (BEM) into power system models in order to assess the potential system value of deploying smart electric thermal storage (SETS) and then to validate them using trial data.

A BEM is a physics-based simulation of building energy use. Inputs into the model include physical characteristics such as building geometry, construction materials, lighting, HVAC[5] and so on (Negendahl 2015; Clarke and Hensen 2015). The model also needs information about building use, occupancy and indoor temperature. A BEM program combines these inputs with information about local weather to calculate thermal loads and energy requirements, the electricity grid's response to those loads and resulting energy use. Such models are used by building professionals and researchers to evaluate the energy performance of buildings for applications like building design, retrofit decision-making, LEED certification[6] and urban planning. Bottom-up models of demand are based on uncertain assumptions (McKenna et al. 2017). To help deal with some of these, the models were initially calibrated based on 'archetypal' data from national databases to allow time to run the simulations required by the project (Andrade-Cabrera et al. 2016). Originally, there was a plan to use trial data at a later stage, to validate the models and recalibrate them if necessary.

5.2.2 Customer Impact Assessment

In parallel with the modelling work, customers were recruited for the live trial and had a combination of technologies installed in their homes, an experience captured in the Customer Impact Assessment (Darby et al.

2018). The technologies installed included heaters and/or hot water cylinders, an internet connection if not already present, a gateway to link the appliances to the cloud (where demand response (DR[7]) would be facilitated), interval meters and, in a sample of homes, additional sensors (occupancy and temperature) and smart plugs[8]. Each home was therefore a source of multiple data points, for assessing the potential for DR and other research purposes.

The social scientists also collected data, including surveys before and after the installation of the technology and at the end of the project, in-home interviews, observations and photographs in a subset of properties and interviews with other project actors (installers, project delivery coordinators, manufacturers, etc.) on their interactions with customers. The objectives were to understand the impact of the installed technologies and, eventually, DR, on customers, and to assess necessary conditions for a good customer experience and DR participation. Five conditions emerged: comfort, control, cost, care and connectivity.

Both the technical and social data were meant to facilitate multidisciplinary collaboration. Interesting data from the implementation phase included indoor and outdoor temperature, occupancy, building fabric, energy consumption (ideally, with heating consumption disaggregated) and customer data held by other partners, like billing, call centre data and DR performance data. The quantitative data from the technologies installed in homes was to be used to triangulate the qualitative data.

5.3 The Processes of Collecting, Sharing and Analysing Data Are Socio-technical

Based on the social and technical contexts just described, researchers took the view that this was a socio-technical project (Foulds and Robison 2017). Following Powells et al. (2014) who argue that electricity 'load' is not an isolated physical phenomenon but also represents activities and social practices, we recognised that the technology and its users were inextricably interlinked and that, therefore, multiple disciplinary methods were necessary. Table 5.1 summarises the data collected.

It also became clear that the *processes* of collecting, sharing and analysing data were socio-technical, no matter whether the data being collected was qualitative or quantitative and irrespective of the use to which it was

Table 5.1 Summary of data collected

Data	Significance for energy (heating) outcomes	Collection methods
Demographics	Age, gender, occupation, education and income level may have an influence on energy use for heating (Wilson and Dowlatabadi 2007)	Some data held by energy company, survey, in-home interviews/observation[a]
Practices	Achieving thermal comfort is not just a matter of an insulated building or efficient heating technologies but also includes skills, meanings and activities (Gram-Hanssen 2013)	Survey, in-home interviews/observation,[a] photographs,[a] installer stories,[a] customer call records,[a] appliance monitoring data (room temperature, comfort settings, boost activity)
Occupancy	The number of people in a house and when they are at home—the assumption in occupancy models is that this is when they use most energy for heating (Richardson 2008; Guerra-Santin and Silvester 2017)[9]	Survey, charging schedules of appliances, interval meters, movement sensors[a]
Consumption	How much electricity is used by the home—as the heating is electric, disaggregated data is important	Survey, billing data, interval metres (for disaggregated data), smart plugs (for particular items)
Building	The size, fabric, age and type of building are key indicators of its energy performance (Gram-Hanssen 2013)	Survey, technical survey, observation,[a] photographs[a]
Temperature	How warm the home is—also a defining indicator of energy use (Peeters et al. 2009)	Core temperature of appliance, appliance sensor, in-home interviews/focus groups,[a] observation,[a] room temperature sensors[a]
Cost	How much customers were spending on their energy consumption and whether this had increased or decreased—ideally a mixture of perceptual and measured data[10]	Survey, in-home interviews/focus groups,[a] bills, actual cost (inserted by agent in final survey)
Appliance monitoring data	Data from heaters (and cylinders)	Thermostatic set point temperature, room temperature, charge period, smart electric thermal storage (SETS) demand request, charge power rate and boost function activity

[a]Only possible in a subset of properties

finally put.[11] For example, the social scientists collected and shared customer satisfaction data with industry partners, building and occupancy data with the modelling team and the interview, observation and photographic data with several partners who were interested in a more in-depth insight into their customers, often to improve the technology on offer. In return, they hoped to receive more quantitative data such as heating periods and temperature settings from the SETS, call centre complaints/inquiries, cost data from the energy providers and consumption data from interval meters.

Having discussed the *use* of the data, we now turn our attention to the data itself. There is no space to deal with every data source in turn but, in the discussion that follows, we explore more fully the idea that dealing with data is socio-technical by focusing on four aspects of the data collection and sharing process necessary to achieve data synergy.

5.4 Data Synergy

We contend that good data depended on four interlinking dimensions:

- Time (synchronising the collection and sharing of data between different parts of the project)
- People (coordinating the different actors involved in the collection and sharing of data)
- Technology (establishing the connectivity between the different technologies so that data could be transmitted)
- Quality (ensuring data is good enough for the research purpose)

The discussion will examine challenges in relation to these dimensions in order to make recommendations for the development of a data protocol for appropriate data synergy for use in other multidisciplinary energy demand projects.

5.4.1 Time: Synchronisation

Figure 5.2 shows the timing of data collection in the project,[12] including the winter periods (critical data collection opportunities in a heating project), the two strands of the project and the variety of data collection methods. It is noteworthy that most data was collected towards the end of the project, with a gap in the middle caused by recruitment difficulties.

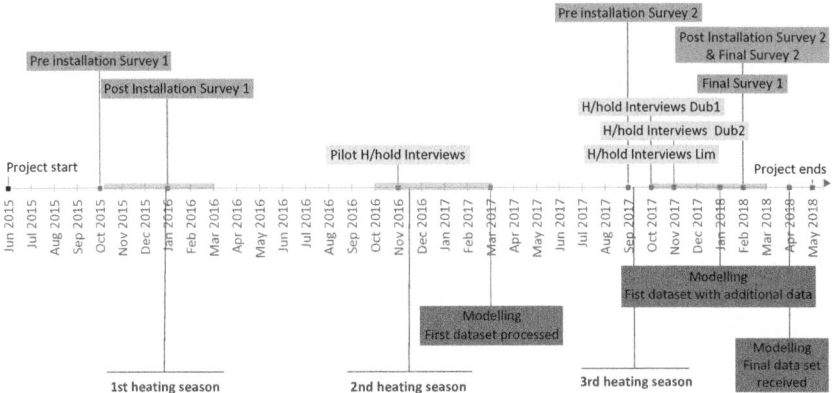

Fig. 5.2 The timing of data collection across the two project strands

Several issues emerged:

1. Multiple data collection methods required complex coordination with the main implementation phases of the project such as recruitment, installation and the three heating seasons, as well as maintaining a coherent approach across the three countries.
2. As different stages started and finished, the need to facilitate communication among actors across different stages of the process became more complicated, and, without a single data person to oversee this process, the inevitable result was that partners focused more on managing their own data and results than on collaboration.
3. Collecting the same data at different points in the project necessitated the altering of the data collection tools to reflect the changing priorities of partners, resulting in changed metrics in some cases, and this compromised the quality of the data and made comparisons across countries difficult.
4. Timing data collection to happen during the winter season was critical, and the ambitious timeframe meant there were only three heating seasons in which to test the technology and monitor behaviour. The first phase of installations had been done by the first heating season, but the connectivity problems discussed below meant data was absent or of poor quality. Further, recruitment was then delayed until just before the final heating season, so close to the end of the project that it was difficult to process data collected when the technologies were at their most reliable.

5.4.2 People: Coordination

People are a crucial part of collecting data, even when the methods are apparently technical. It is worth noting the different roles of people in the project, each of whom impacted the data: customers, data collection agents, installers, industrial project partners and researchers. The nature of this project meant direct access to customers was restricted, and so data were generally not collected by researchers. This was problematic because those collecting it did not have the skills, training or appreciation of the final use of the data to collect it correctly, as they had other priorities.

Previous research (Janda and Parag 2013; Wade et al. 2016) has highlighted the influence of different actors in socio-technical processes, and this project was a case in point. The otherwise excellent project management team had an industry background, and their priority was implementation rather than research. Thus, ensuring timely deliverables sometimes hampered the collection and sharing of research data. Table 5.2 serves to highlight the number of different actors involved in the project and consequent complexity of sharing different types of data.

Apart from the logistical challenge of coordinating the data across actors, working with multiple partners had other challenges, more widely discussed in the literature, such as a lack of shared ontology, vocabulary and culture (Hargreaves and Burgess 2009; Longhurst and Chilvers 2012; Robison and Foulds 2017; Sovacool et al. 2015). Data sets also had a different meaning for different partners, who brought different skills to the analysis and interpreted, and then used, the data differently. This had implications for the quality of data they needed and the way in which the data was interpreted, both of which are discussed later under data quality.

5.4.3 Technology: Connectivity

> Given IOT [Internet of Things] is in the news… clean technology, all these buzzwords are always being used. But yet, when it comes to the practicalities of doing a project with [hundreds of] houses, it was incredibly difficult.
>
> Project delivery coordinator, RealValue project

Good connectivity between the different technologies was essential, both for successful DR and to access most of the quantitative data. It is not necessary to dwell on the details of these connections (Fig. 5.3), but, in essence, it was necessary for the connected appliances to communicate

Table 5.2 Actors involved in collecting and sharing different types of data sets

Types of data sets	Actors in data collection/sharing						
	Energy suppliers	Manufacturers	Software and hardware developers	Energy networks operators	Internet providers	Installers	Research organisations
General customer contact details	✓	✓	✓		✓	✓	✓
Customer communications (calls, emails)	✓						
Customer surveys[a]	✓	✓					✓
Customer interviews	✓	✓					✓
Pre-installation data (technical survey)	✓					✓	
Installation data (no. of heaters, cylinder/gateway, etc.)	✓	✓	✓	✓	✓	✓	✓
Demand response (DR) performance data	✓	✓	✓		✓		✓
Participant contract	✓						
Metering data	✓	✓		✓	✓		
3G usage data	✓		✓				

[a]In some cases an agency carried out customer surveys

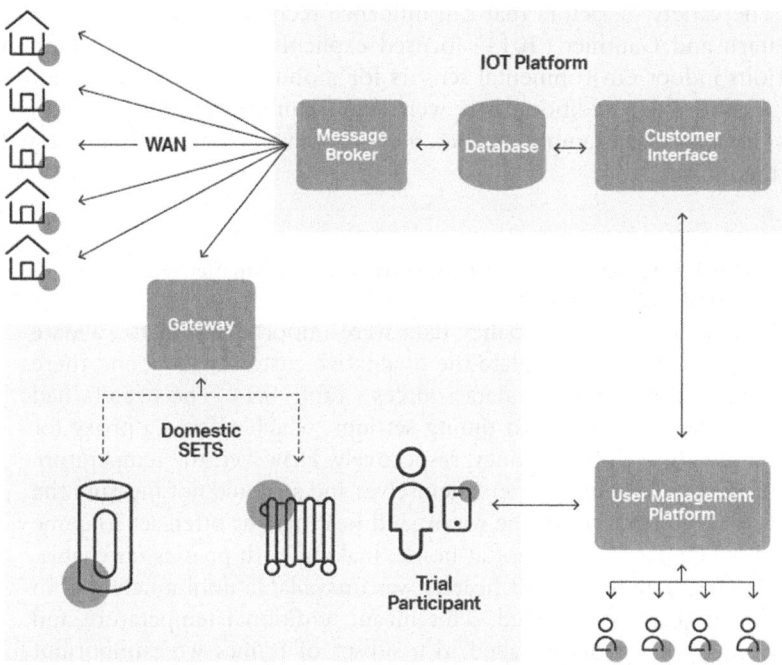

Fig. 5.3 Diagram of the subsystem integration and data flows behind the RealValue user interface application. WAN = Wireless Area Network, IoT = Internet of Things, SETS = Smart Electric Thermal Storage. Source: RealValue project partners, cited in Darby et al. (2018)

through a gateway to a cloud-based aggregation platform that optimised the charging of those appliances according to the customer's comfort settings, cost algorithms and grid constraints. This was unexpectedly demanding. Unanticipated complications included the need to install internet connections, customers turning off one or other technology, power failures causing the appliances to revert to 'stand-alone' mode (i.e. not connected and so no longer transmitting data or available for DR), the need to develop interfaces for different technologies to communicate, organisational firewalls preventing communication, changing communication protocols necessitating ongoing modifications and a software update that disrupted the appliances.

The variety of factors that can influence technical data is noteworthy. Spataru and Gauthier (2014) focused explicitly on the performance of various indoor environmental sensors for monitoring people and indoor temperatures. In addition, there were significant impacts on the researchers (for a specific example, see Box 5.1). However, we are more interested in the impact.

> **Box 5.1 Attempts to collect temperature and occupancy data using technical and social methods**
>
> Temperature and occupancy data were important both to validate the models and triangulate the qualitative customer data, and there were multiple possible data sources (Table 5.1). The heaters had temperature sensors and timing settings, which offered a proxy for temperature and occupancy, respectively. However, the temperature sensors were on the heaters themselves and so could not measure the actual temperature of the room, and heating was often set to come on when people were not at home, making both proxies unreliable. Besides, data from most heaters was unavailable until much later in the project, as described. This meant additional temperature and occupancy sensors installed in a subset of homes were important both to help calibrate the models with this appliance data and to triangulate the customer impact assessment data, but there were two significant problems. The first was that most did not transmit data. The second was that the location of the sensors was not accurately noted by those who installed them, making interpretation of the data impossible.
>
> Although the social scientists included occupancy and temperature questions in the surveys, these were filled in by agents with different objectives, and the data was incomplete and ultimately unusable. Follow-up home visits were carried out and did include questions and observations on temperature and occupancy that were shared with modellers, but it was not possible to visit the homes with additional sensors, again because of the need to coordinate with other project partners, and so remedying the connectivity issues or observing the location of the sensors was impossible. Despite multiple possible sources, therefore, the final data on temperature and occupancy was patchy. This prevented researchers collaborating as fully as they might have done otherwise.

5.4.4 Data Quality: Granularity, Reliability and Project Design

During the final heating season, recruitment was completed and attention turned to fixing the connectivity issues, with some success: data did become available. As partners started to work with it, however, the next major issue arose—the quality of the data, a product of the previous three sections (Stevenson and Leaman 2010). All sorts of factors had affected the data but there are three main points to discuss here.

First, expectations of the granularity (or resolution[13]), and the duration of the data, varied depending on the partner and their purpose. So, whilst industrial partners needed single 24-hour periods of uninterrupted data to run equipment diagnostics, social scientists wanted data for participants for whom they had other data (such as surveys or interviews), and modellers needed several days of data to help them see patterns but did not mind some gaps, as long as they had an idea of occupancy (Fig. 5.4).

Second is the reliability, or consistency, of the data. As noted in Fig. 5.4, different methods of collecting apparently the same data yielded different results, making methodological transparency and accuracy vital for replicable research. Figure 5.5 demonstrates this from viewpoint of the data. It shows two sets of temperature data: one from a SETS temperature sensor, the other from an additional temperature sensor (whose location was unknown).

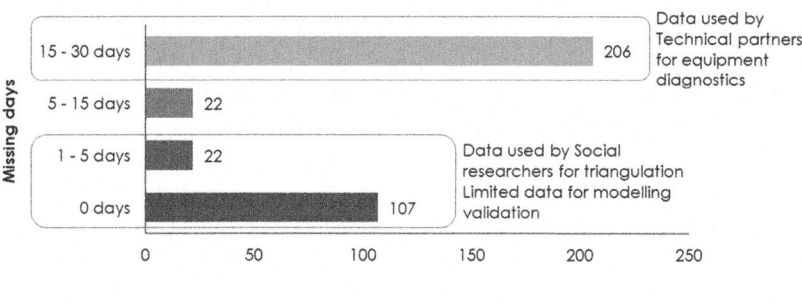

Fig. 5.4 Missing days of SETS monitoring data from Irish data sample ($n = 357$) in September 2017

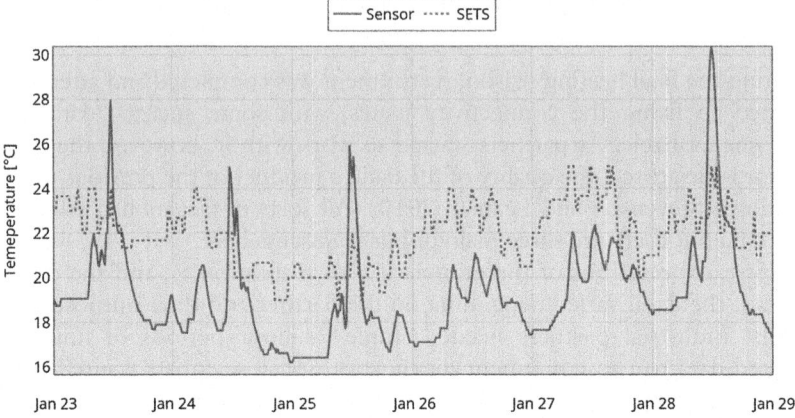

Fig. 5.5 Sensor temperature data comparison in a single household in 2018 (potential misplacement)

Based on the midday temperature spikes on the solid line, we could speculate that the additional sensor was warmed by the sun. Interestingly, the interpretation of what happened on the days without spikes differed between modellers and social scientists: the former assuming cloudy weather and the latter closed curtains, possibly indicating illness or shift work, for example. Without additional data on weather, the aspect of the room and occupancy, it is not possible to tell which of these is correct, but the different analyses indicate each discipline's bias.

Still on temperature, the 2–4 °C difference between the two sensors is striking.[14] As the SETS sensor is on the metallic SETS surface near the warm air vent, it might well be warmer than the room. This might help explain the high temperature settings seen during the home visits: 24 °C at the appliance might translate to 18–20 °C in the room.

Both graphs also show gaps in the data, indicated by straight horizontal lines. Strangely, these do not always coincide, suggesting either that they were caused by different factors or that there were various combinations of factors affecting data quality. Again, without a home visit to verify, the cause cannot be known.

Third is the socio-technical project design. What has become clear upon examination of the data is that many of the problems related to the project design phase of the project. Rather than a socio-technical project design, this was in fact an industry-led technical demonstration project with some social inputs, partly leading to the incommensurability of the data discussed above. A socio-technical project design should encompass three phases: model, design and methods, and analysis, all of which should be socio-technical. This should start with a conceptual, theoretical phase that considers how the actions and states of people interact with the technical and physical properties of their environments. It might end with an analysis of socio-technical constructs such as a 'person-space-time mean internal temperature', a measure meant to get closer to the user experience of temperature in the home (Love and Cooper 2015). The methods linking these have yet to be developed, but mobile phones and in-home temperature apps might offer some traction (Grunewald 2015).

5.5 Achieving Data Synergy

Epistemological debates run as an undercurrent through all of these issues. Fundamentally, the more positivist-grounded technical/monitoring sciences would define quality in very different ways to most critical social scientists, who would instead embrace subjectivity, implying that issues of 'validation' and 'calibration', in the traditional sense, are backgrounded or at least mean something different. Nevertheless, in the context of a replicability crisis in various disciplines, this chapter suggests that data processes in the energy demand research community could use improvement.

We have contributed to the conversation about ways in which this might happen and will finish with recommendations in each of the four dimensions discussed:

- **Time:** Synchronising research rests on critical dependencies, different from project management, and requires backup plans to ensure quality data, otherwise sometimes constrained by the project plan. Also, the duration of heating projects needs to be better aligned with their objectives.[15]
- **People:** The impact of different actors cannot be underestimated. Planning and responsive management are essential parts of real-world project delivery, and we would recommend four coordination roles—a project manager, a project delivery coordinator (for practical project implementation), a data analyst (from the start of the project, to organise, hold and facilitate access to a shared set of data) and a research coordinator (with a socio-technical background, to synchronise the research).

- **Technology:** Demonstration projects inevitably use novel technologies and the difficulty of managing the interfaces between them should be taken into account.
- **Quality:** The use of consistent metrics would allow better comparisons across different countries with different languages, contexts, technologies and participant groups. Data protocols need to be developed to establish conventions for collecting and sharing data, both quantitative (e.g. what to capture, how often and where) and qualitative (e.g. what scales to use for age, income and cost).

This is not trivial and requires work from researchers and funders. However, the reward would be more robust, reliable data; better, more policy-relevant outcomes; and more replicable research.

Acknowledgements The RealValue project was funded by the European Union's Horizon 2020 research and innovation programme under grant agreement no. 646116. We would like to thank our participants, without whom the research would not be possible, and the reviewers, whose insights vastly improved our chapter.

Notes

1. The Oxford Dictionary defines multidisciplinary as 'Combining or involving several academic disciplines or professional specializations in an approach to a topic or problem'. This fits our purposes in this chapter.
2. Such as the need for larger/representative/standardised samples vs. the need for depth/bringing out individual differences in the data, for example.
3. Data synergy is a term coined for this chapter and describes data from multiple sources or disciplines that, when combined, is more valuable than any of the sources were on their own.
4. http://www.realvalueproject.com.
5. HVAC (heating, ventilation and air conditioning).
6. LEED (Leadership in Energy and Environmental Design) is the most widely used green building rating system in the world.
7. Demand response seeks to adjust the demand for power instead of adjusting the supply, for the benefit of the grid.
8. A plug that provides control of any device plugged into it.
9. Though this may not be true where indoor temperatures are kept constant using thermostats and where, in fact, people warming the environment through their bodies and activities may lessen the need for heating.
10. In many countries the weather varies from one year to another, and thus the heat demand. In fact, therefore, information on heating costs at stable

tariffs in a normal year are required (the effect of price increases and weather variations should be discounted).
11. Love and Cooper (2015) discuss the need for socio-technical data rather than separate streams of social and technical data.
12. This is just the timeline for Ireland. Data was also collected in Latvia and Germany.
13. The number of data points within a particular period for a particular data set.
14. Higher variations than this have been recorded. To make some kind of judgement here, one needs data from many homes and sensors, but only 5 of 50 homes installed with additional room temperature sensors provided usable data, and even this was not good quality.
15. This is out of the control of the project itself but something that should be considered by funders.

References

Andrade-Cabrera, C., Turner, W. J. N., Burke, D., Neu, O., & Finn, D. P. (2016). Lumped Parameter Building Model Calibration using Particle Swarm Optimization. *Proceedings of ASIM 2016: 3rd Asia Conference of International Building Performance Simulation Association*, Jeju, South Korea.

Clarke, J. A., & Hensen, J. L. M. (2015). Integrated Building Performance Simulation: Progress, Prospects and Requirements. *Building and Environment, 91*, 294–306.

Darby, S., Higginson, S., Topouzi, M., Goodhew, J., & Reiss, S. (2018). *Getting the Balance Right. Can Smart Electric Thermal Storage Work for Both Customers and Grids? Report for the RealValue Project (Deliverable 5.3)*. Oxford: Environmental Change Institute.

Economidou, M., Atanasiu, B., Despret, C., Maio, J., Nolte, I., & Rapf, O. (2011). *Europe's Buildings Under the Microscope: A Country-by-country Review of the Energy Performance of Buildings*. Brussels: Buildings Performance Institute Europe (BPIE).

Foulds, C., & Robison, R. (2017). *The SHAPE ENERGY Lexicon—Interpreting Energy-related Social Sciences and Humanities Terminology*. Cambridge: SHAPE ENERGY.

Gram-Hanssen, K. (2013). Efficient Technologies or User Behaviour, Which Is the More Important When Reducing Households' Energy Consumption? *Energy Efficiency, 6*, 447–457.

Grunewald, P. (2015). Measuring and Evaluating Time-Use and Electricity-Use Relationships (Meter). Early Career Fellowship Ref. EP/M024652/1. *Engineering and Physical Sciences Research Council EPSRC)*.

Guerra-Santin, O., & Silvester, S. (2017). Development of Dutch Occupancy and Heating Profiles for Building Simulation. *Building Research and Information, 45*(4), 396–413.

Guerra-Santin, O., Tweed, C., Jenkins, H., & Jiang, S. (2013). Monitoring the Performance of Low Energy Dwellings: Two UK Case Studies. *Energy and Buildings*, *64*, 32–40.

Gupta, R., & Kapsali, M. (2015). Empirical Assessment of Indoor Air Quality and Overheating in Low-carbon Social Housing Dwellings in England, UK. *Advances in Building Energy Research*, pp. 1–23.

Hargreaves, T., & Burgess, J. (2009). *Pathways to Interdisciplinarity: A Technical Report Exploring Collaborative Interdisciplinary Working in the Transition Pathways Consortium*. Norwich: School of Environmental Sciences, University of East Anglia.

Janda, K. B., & Parag, Y. (2013). A Middle-out Approach for Improving Energy Performance in Buildings. *Building Research and Information*, *41*, 39–50.

Jones, P., Lannon, S., & Patterson, J. (2013). Retrofitting Existing Housing: How Far, How Much? *Building Research and Information*, *41*, 532–550.

Longhurst, N., & Chilvers, J. (2012). *Interdisciplinarity in Transition? A Technical Report on the Interdisciplinarity of the Transition Pathways to a Low Carbon Economy Consortium*. Norwich: Science, Society and Sustainability, University of East Anglia.

Love, J., & Cooper, A. (2015). From Social and Technical to Socio-technical: Designing Integrated Research on Domestic Energy Use. *Indoor and Built Environment*, *24*(7), 986–998.

Mallaband, B., Wood, G., Buchanan, K., Staddon, S., Mogles, N. M., & Gabe-Thomas, E. (2017). The Reality of Cross-disciplinary Energy Research in the United Kingdom: A Social Science Perspective. *Energy Research & Social Science*, *25*, 9–18.

McKenna, E., Higginson, S., Grünewald, P., & Darby, S. J. (2017). Simulating Residential Demand Response: Improving Socio-technical Assumptions in Activity-based Models of Energy Demand. *Energy Efficiency*, pp. 1–15, doi:https://doi.org/10.1007/s12053-017-9525-4.

Negendahl, K. (2015). Building Performance Simulation in the Early Design Stage: An Introduction to Integrated Dynamic Models. *Automation in Construction*, *54*, 39–53.

Peeters, L., de Dear, R., Hensen, J., & d'Haeseleer, W. (2009). Thermal Comfort in Residential Buildings: Comfort Values and Scales for Building Energy Simulation. *Applied Energy*, *86*(5), 772–780.

Powells, G., Bulkeley, H., Bell, S., & Judson, E. (2014). Peak Electricity Demand and the Flexibility of Everyday Life. *Geoforum*, *55*, 43–52.

Richardson, I. (2008). A High-resolution Domestic Building Occupancy Model for Energy Demand Simulations. *Energy and Buildings*, *40*(8), 1560–1566.

Robison, R., & Foulds, C. (2017). Creating an Interdisciplinary Energy Lexicon: Working with Terminology Differences in Support of Better Energy Policy, Proceedings of the eceee 2017 Summer Study on *Consumption, Efficiency & Limits*, paper 1-267-17. 29 May–3 June 2017, Presqu'ile de Giens, France. pp. 121–130.

Sovacool, B. K., Ryan, S. E., Stern, P. C., Janda, K., Rochlin, G., Spreng, D., Pasqualetti, M. J., Wilhite, H., & Lutzenhiser, L. (2015). Integrating Social Science in Energy Research. *Energy Research & Social Science, 6,* 95–99.

Spataru, C., & Gauthier, S. (2014). How to Monitor People 'Smartly' to Help Reducing Energy Consumption in Buildings? *Architectural Engineering and Design Management, 10,* 60–78.

Stevenson, F., & Leaman, A. (2010). Evaluating Housing Performance in Relation to Human Behaviour: New Challenges. *Building Research & Information, 38,* 437–441.

Topouzi, M., Grunewald, P., Gershuny, J., & Harms, T. (2016). Everyday Household Practices and Electricity Use: Early Findings form a Mixed-method Approach to Assessing Demand Flexibility. Proceedings of the *BEHAVE 2016 4th European Conference on Behaviour and Energy Efficiency.* 8–9 September 2016, Coimbra, Portugal.

TSB. (2014). *Retrofit for the Future: Reducing Energy Use in Existing Homes. A Guide to Making Retrofit Work.* Swindon: Technology Strategy Board.

Wade, F., Hitchings, R., & Shipworth, M. (2016). Understanding the Missing Middlemen of Domestic Heating: Installers as a Community of Professional Practice in the United Kingdom. *Energy Research & Social Science, 19,* 39–47.

Wilson, C., & Dowlatabadi, H. (2007). Models of Decision Making and Residential Energy Use. *Annual Review of Environment and Resources, 32,* 169–203.

Open Access This chapter is licensed under the terms of the Creative Commons Attribution 4.0 International License (http://creativecommons.org/licenses/by/4.0/), which permits use, sharing, adaptation, distribution and reproduction in any medium or format, as long as you give appropriate credit to the original author(s) and the source, provide a link to the Creative Commons license and indicate if changes were made.

The images or other third party material in this chapter are included in the chapter's Creative Commons license, unless indicated otherwise in a credit line to the material. If material is not included in the chapter's Creative Commons license and your intended use is not permitted by statutory regulation or exceeds the permitted use, you will need to obtain permission directly from the copyright holder.

CHAPTER 6

Building Governance and Energy Efficiency: Mapping the Interdisciplinary Challenge

Frankie McCarthy, Susan Bright, and Tina Fawcett

Abstract Improving the energy efficiency of multi-owned properties (MoPs)—commonly known as apartment or condominium buildings—is central to the achievement of European energy targets. However, little work to date has focused on how to facilitate retrofit in this context. Drawing on interdisciplinary Social Sciences and Humanities expertise in academia, policy and practice, this chapter posits that decision-making processes within MoPs might provide a key to the retrofit challenge. Existing theories or models of decision-making, applied in the MoP context, might help to explain how collective retrofit decisions are taken—or overlooked. Insights from case studies and practitioners are also key.

F. McCarthy (✉)
School of Law, University of Glasgow, Glasgow, UK
e-mail: frankie.mccarthy@glasgow.ac.uk

S. Bright
Centre for Socio-Legal Studies, University of Oxford, Oxford, UK
e-mail: susan.bright@law.ox.ac.uk

T. Fawcett
Environmental Change Institute, University of Oxford, Oxford, UK
e-mail: tina.fawcett@eci.ox.ac.uk

© The Author(s) 2018
C. Foulds, R. Robison (eds.), *Advancing Energy Policy*,
https://doi.org/10.1007/978-3-319-99097-2_6

Theories of change might then be employed to develop strategies to facilitate positive retrofit decisions. The chapter maps the issues and sets an agenda for further interdisciplinary research in this novel area.

Keywords Energy efficiency • Apartments • Condominiums • Collective action • Governance

6.1 Context

Around 40% of European citizens live in multi-owned (apartment or condominium) buildings (Bright and Weatherall 2017). Improvements to the energy efficiency of multi-owned properties (MoPs) and the energy behaviour of residents are therefore essential to the achievement of Europe's energy goals. Existing work on energy efficiency in the housing context tends to focus on single-family dwellings, ignoring the additional complexities which arise where the participation of multiple parties is required (Matschoss et al. 2013; Weatherall et al. 2017). The Governance and Renewable Energy in Efficient Apartments Network for the European Union (GREEAN-EU) is an interdisciplinary research network of researchers, policymakers and practitioners that was formed to address this gap in the academic and policy discussion. The focus of the network is on how the opportunities for energy efficiency and upgrades in MoPs are affected by building governance (as explained in Bright and Weatherall 2017). This requires an understanding of 'the technology of law', that is, how in different European jurisdictions MoP laws structure decisions about the use of energy and energy-related technologies ('energy decisions'). This will depend on country-specific legal rules, practices, title and ownership arrangements used to regulate management of MoPs, as well as the way in which law mediates and structures decision-making. Energy decisions are impacted not only by these legal considerations but also by organisational factors, that is, how human actors work as a decision-making community, yet little attention has previously been given to understanding this in the context of MoPs.

This chapter reports on discussions from a workshop in Oxford in March 2018. The workshop's objective was to develop a conversation around new research approaches for understanding MoP energy-related decision-making processes by bringing together both legal and organisational aspects of building governance. The workshop was exploratory, with participants who covered a broad range of disciplines within the Social Sciences and Humanities. As well as experts in property law and energy policy and practice, most of whom were already familiar with our research questions, it included researchers in Psychology, Sociology and Economics with expertise in group decision-making, but not necessarily in relation to the European energy agenda or MoPs. Participants were invited based on knowledge of their existing work and its relevance to our research agenda. Whilst discussions were inevitably partial and selective, representing the perspectives of the (sub)disciplines represented at the workshop, this chapter offers substantive insights on how disciplinary perspectives can be integrated to provide answers to this critical research question and practical insights on how to conduct this work effectively. In line with the aims of the GREEAN-EU network, this chapter sets out the need for solutions (to the challenges presented by this critical but overlooked area within the European energy field) to be developed by working across disciplinary boundaries inside and outside the academic world. By doing so, it will contribute to new ideas on how the governance barrier to energy efficiency can be reduced, removed or transformed into a positive driver.

6.2 Building Governance and Energy Efficiency: Key Research Questions

Article 19 of the 2012 Energy Efficiency Directive (2012/27/EU) requires member states to take appropriate measures to remove regulatory and non-regulatory barriers to energy efficiency in MoPs. Property law is central to this obligation, and yet its role has gone largely unexamined (Bright and Weatherall 2017). Bright and Weatherall began addressing this key aspect of energy behaviour through their Futureproofing Flats project, which provided a basis for the development of the 'Building Governance Model' (Bright and Weatherall 2017), a new framework within which to conduct an investigation in a range of European jurisdictions of the challenges of MoP retrofit. Its starting point was the literature on the 'energy efficiency gap' between actual energy use and optimal levels of energy efficiency (Hirst and Brown 1990; Jaffe and Stavins 1994), which includes a range of barrier models explaining the existence of this gap in particular contexts (Lutzenhiser 1993; Sorrell et al. 2011; Janda et al. 2015). Within that literature, some recent work has focused on the barriers specific to MoPs (Matschoss et al. 2013; LEAF 2016). The LEAF project, for example, grouped the challenges for multi-family properties into four categories: technical issues, agreement issues, financial issues and the behaviour of residents. Drawing on this research, the Building Governance Model posits governance as a type of meta-category which both represents and shapes the barriers described in earlier models. The structure within which energy decisions can be taken in MoPs is delineated by a combination of the law of property and the law of associations. Property law rules as to who owns which parts of a MoP building (such as the roof, windows and foundations) play a role in determining who, if anyone, has the power to instruct retrofit work. The law of associations structures how decisions involving multiple owners or residents can be taken, including meeting arrangements, voting processes and decision-making thresholds. A combination of these rules will determine who holds responsibility for the costs of the work, and whether and how finance can be accessed. The complexity of the law, in conjunction with regulation determining the availability of data on the energy efficiency of the building and the range of improvements possible, can create significant informational challenges for decision-makers. Law, therefore, creates a unique set of challenges to MoP energy efficiency.

The Building Governance Model draws attention to two further areas for investigation that require interdisciplinary collaboration between academics across the Social Sciences and Humanities, as well as with legal and energy practitioners and experts from the policy community. The first research area concerns gathering the data necessary to understand the full scope of the governance challenge. To address this, a cross-European group of property law researchers and energy policy practitioners within GREEAN-EU are developing a set of interdisciplinary methodologies for collection and analysis of data on relevant legal frameworks, building stock, energy use and energy performance in MoPs. The second research area concerns the process of energy-related decision-making within MoPs. The central ambition here is to develop a framework within which to understand how complex groups may be able to take energy decisions that benefit them collectively as well as individually. This chapter draws on the expertise represented at the Oxford workshop in order to explore how different disciplinary insights can contribute to that framework.

6.3 Energy Decision-Making in MoPs: Issues for Further Research

The workshop provided an opportunity for all participants to better understand GREEAN-EU's research questions and to consider how expertise from their respective fields could be brought together to address those questions. The day entailed (i) an overview of GREEAN-EU's work, (ii) a review of earlier related projects carried out by attendees, (iii) a 'teach-in' where attendees outlined the contribution their discipline might make to addressing the decision-making question and (iv) a 'hackathon' where combinations of attendees explored how they might collaborate to take this work forward.

The issues around MoP energy-related decision-making were mapped within two broad questions. First, how are energy-related decisions taken in MoPs? This focuses on describing existing processes and identifying the reasons for them. Second, how can this understanding of decision-making processes be used to accelerate the rate of energy renovations to MoP buildings? This focuses on understanding how changes to the legal, social and economic context within which decisions are made might lead to better energy outcomes. Drawing on the cross-disciplinary academic and practitioner expertise represented at the workshop, we identified a range of possible approaches to addressing these questions and outlined the further research that would be required to develop a complete interdisciplinary methodology here.

The workshop also revealed potential gaps in our disciplinary coverage. On the academic side, we concluded that collaboration with researchers in Anthropology and Human Geography might offer useful insights into decision-making. On the practitioner side, expertise from the fields of investment/financing, communication and group mediation would be beneficial in developing strategies for changing the outcomes of decision-making processes.

6.3.1 How Are Energy-Related Decisions Taken in MoPs?

6.3.1.1 Understanding Collective Decision-Making

Various models of decision-making within the Social Sciences were discussed which may be useful in understanding decision-making in apartment blocks. This was not a comprehensive discussion of all models of human choices, actions or practices, and theories which decentre the individual, such as Social Practice Theory, are not represented here. Some models, such as Collective Efficacy Theory developed by sociologist Robert Sampson in the context of controlling crime (Sampson et al. 1997), focus on the importance of wider groups in activating social ties to achieve collective goals and examine the contextual factors which support or obstruct that control. Others focus more upon the importance of wider groups in individual choices. The Theory of Planned Behaviour (Ajzen 1991), as used in Social Psychology, considers that decisions result from three sets of beliefs held by the decision-makers: about the likely consequences of decisions (attitudes), about the normative expectations of others inside and outside the group in respect of the decision (subjective norms) and about factors which may support or obstruct the performance of the decision (controls). Although this theory has been very influential in understanding decisions, some commentators suggest that it does not help practitioners to develop helpful interventions (Sniehotta et al. 2014). Social Identity Theory (Tajfel and Turner 1986) seeks to explain the actions of individuals by reference to the groups to which they consider themselves to belong. Thus, behaviours or decisions of the individual may be influenced by others who they see as fellow 'in-group members'. On the other hand, the Bystander Intervention Model (Darley and Latane 1968) predicts that under conditions of ambiguity of responsibility or where there is perceived diffusion of responsibility, individuals may fail to act (e.g. to improve energy efficiency).

A key research question is whether these theories can be usefully adapted to the MoP context. The ACE Retrofitting (2018) project, through which local governments aim to accelerate the energy retrofitting of condominiums by acting as facilitators between co-owners and building professionals, makes use of the Theory of Planned Behaviour in its work on development of tools for accelerating change. But there is a need for more Social Science research to understand how energy decision-making is best explained in the MoP context and what can be learned from this to support better energy outcomes.

6.3.1.2 Forming the Collective
The models described above aim to understand the behaviours of a group once it has formed. In MoPs, however, it is not necessarily the case that the individuals with power to take retrofit decisions perceive themselves as

a group or act in a collective manner. In legal systems like Scotland and England, ownership of part of a MoP does not carry with it any obligation to meet with fellow owners or to establish a body (like an owners' association) to act in their collective interests; apartment owners may perceive themselves simply as individuals who happen to share a building (Weatherall et al. 2017). In legal systems such as France—where establishment of an owners' association and employment of a property manager are mandatory—although owners may understand themselves to be part of a collective, they may not consider themselves to have *responsibility* for the collective, viewing that instead as the role of the manager.

Addressing this aspect of MoP behaviour requires an understanding of how groups come to be formed. However, there appears to be no global theory within any of the disciplines represented in GREEAN-EU that can help us to understand this process. Further research is needed to fill this gap.

6.3.1.3 Use Decisions and Investment Decisions

Our workshop discussions suggested that decisions regarding energy can be broadly split into use and investment decisions. Use decisions, such as what temperature homes are heated to, are more usually conceptualised as habits, behaviours or practices. Investment decisions, including energy renovation and retrofit works, require a conscious decision to invest. It is these latter decisions which form the focus of GREEAN-EU's work.

There is limited theoretical material to draw on in understanding this form of decision-making outside the field of Economics, where familiar models such as cost/benefit analysis tend to dominate. An underlying presumption in most economic models is that decisions are taken by an economically rational individual who has access to full information, has consistent and stable preferences, and only seeks to maximise their own expected utility. Critics have long argued that this presumption has little basis in reality (Raworth 2017), which has led to the development of new 'behavioural' models of decision-making (e.g. Nudge Theory, explained below). Perhaps more problematically for our research, a model which focuses on the actions of individuals is unlikely to capture the necessary collective aspect of the investment behaviour with which we are concerned. Again, further research would be needed to fill this gap.

6.3.1.4 One Size Does Not Fit All

A final concern relates to the heterogeneity of MoP collectives. The number of individuals involved may range from a minimum of two up to a maximum of several thousand people. As previously discussed, the ownership and management structures will vary from country to country based on domestic legal rules and practices. Cultural context is also likely to play a role in how a group functions. It is important to be aware of the risk of reductionism in attempting to explain the behaviour of every MoP collective by reference to a single model. Retaining an understanding of the differences between groups based on scale, organisation or other factors may be critical to the development of effective solutions.

6.3.2 How Can Understanding MoP Energy Decision-Making Help Accelerate the Rate of Energy Retrofit?

6.3.2.1 Theories of Change

In addition to awareness of how energy decisions are taken in the MoP setting, it is also important to consider how decision outcomes can be aligned with the ambitions of the European Energy Union. Explaining collective action, that is, how groups take action to achieve a common objective when there may be misalignments between individual and collective incentives, has been a challenge within several branches of the Social Sciences. Amongst various theories of change that may be apposite to the MoP energy context, both Nudge Theory and Social Marketing approaches were discussed at the workshop, whilst noting that the interaction between the role of the individual owner and the collective introduces complexity. Nudge Theory considers that suggestion or 'choice architecture' may be the most effective way of influencing decision-making behaviour. A policy tool of 'green nudges' is beginning to emerge which recognises specific issues in the environmental context, for example, the non-economic nature of perceived benefits from energy efficient behaviour, and the challenges of persuading individuals who are sceptical about the existence of climate change of any benefits to this behaviour. A Social Marketing approach concentrates on identifying the barriers to energy efficient behaviour or decisions, such as bystander effect factors and mental models/folk theories (incorrect but tenacious assumptions about the type of behaviour which is efficient), and employing techniques to overcome them. However, these theories or change mechanisms have been developed primarily in the context of habitual behaviour change, rather than investment decisions.

Having a 'champion' within the group to spearhead collective action can be powerful. Such leaders may be viewed as 'spokespersons' of energy transition from the perspective of the Actor Network Theory (Akrich et al. 2006) or as innovative 'in-group' members from the perspective of Social Identity Theory. Case study evidence suggests a champion can play a critical role in relation to MoPs where energy renovations *have* been undertaken (Brisepierre 2011; Le Garrec 2014). Empirical sociological research in France has found that energy improvement decisions in MoPs are often initiated by leaders or groups of leaders living within the building who persuade the wider community to take action. This confirms previous observations made by sociologists who studied French condominiums with the strategic analysis tools provided by the Sociology of Organised Action (Crozier and Friedberg 1977; Golovtchenko 1998; Lefeuvre 1999). From these studies, it is clear that collective action in a condominium depends on the skills of the actors and the capacity for building consensus.

In taking this aspect of our research agenda forwards, a first step would be to identify more case study examples of MoPs in which retrofits have been carried out in order to identify the mechanisms at play in the decision-making process. This research would need to be open to a wide range of theoretical frameworks, not necessarily just those discussed in our workshop.

6.3.2.2 Levers for Change

GREEAN-EU's focus on governance has the result that changes to the law may tend to be foregrounded in our development of proposals for change. However, it is important to avoid becoming blinkered in this respect. A number of practical and policy levers could be utilised, as identified through practitioner and case study experience, as well as from theoretical insights. In particular, the role of actors outside the MoP collective, such as property managers and building professionals (architects, contractors, etc.), could be critical. French sociological research has also highlighted the importance of neutral advice provided by local energy agencies, energy information desks or consumer associations supporting MoP representatives. The role of cities and the potential for them to drive action in relation to retrofit should also be kept in mind. The availability and regulation of finance for retrofit work are also important. Most fundamentally, the availability and accessibility of information about the types of energy renovation possible in different MoP buildings and the legal and financial measures necessary to undertake the work must meet the needs of MoP collectives, or it is unlikely that any progress can be made.

6.4 Energy Decision-Making in MoPs: Practical Challenges of Further Research

In addition to the considerable difficulties with integrating different research approaches within the Social Sciences and Humanities (one example being multiple understandings of the 'active consumer'—Fox et al. (2017)), there are also practical challenges with conducting pan-European interdisciplinary research. These may sound marginal, but discussion in our workshop suggested they present real barriers to collaborative work. A summary of the principal challenges we identified, broken down into key aspects along with suggested solutions, is set out in Table 6.1. Awareness of these issues at an early stage is likely to prove essential to further research in this area.

Table 6.1 Challenges of interdisciplinary research

Challenge	Key aspects	Suggested solutions
Levels of interdisciplinary expertise	• Interdisciplinary work more common in some disciplines than others • Historically rare within law, the central focus of our work • Work with non-academics in policy and practice present different challenges	• Share biographies amongst project team to foster awareness of experience levels • Discuss issue at first team meeting and suggest reading as necessary • Encourage questions

(continued)

Table 6.1 continued

Challenge	Key aspects	Suggested solutions
Variation in disciplinary styles and methods	• Methodological expertise unlikely to be shared by all, and 'shared' methodologies (e.g. theoretical analysis, historical analysis) may have different meanings in different disciplines • Variation in conventions of writing (active/passive voice, article length, approaches to referencing) between disciplines and between academics and non-academics • Identification of appropriate journals/outlets for publication	• Ensure awareness of issue • Basic explanations of key methodologies by team members at early meeting • Develop 'style guide' for writing • Create regular opportunities for discussion of issues as they arise
Language barrier	• Language used for project will not be first language of all team members • Technical terms may have different meanings in different disciplines • Legal terms may vary by jurisdiction	• Ensure awareness of issue • Develop 'glossary' of key project terms
Communication	• Varying expectations of working hours by discipline and country • Identifying appropriate technology for virtual team meetings • Funding for in-person meetings • Most effective structure for meetings	• Discuss issue at first team meeting and develop protocol covering these issues
Management	• Manager(s) must maintain overview of whole project • Keep up with developments across disciplines and in European energy policy	• Ensure time is factored in for management • Build in structures for regular communication to and from manager(s) re: developments

6.5 Next Steps

By building a broader multi-disciplinary network, and mapping key research questions, this work has laid the foundations for the next stage of the GREEAN-EU project. Several theories of change have been identified as potentially relevant to the decision-making question, both in terms of understanding current behaviour and in developing tools for changing that behaviour.

Subject to securing funding, GREEAN-EU aims to develop the research down two connected pathways. One is a comprehensive desk-based exploration of the applicability of the identified theories in the MoP decision-making context. The other is identification of a number of case study MoPs in order to carry out empirical investigation of their experience of retrofit decision-making, testing the application of these theories of change in practice. This work should provide the data necessary for development of a methodology for a more comprehensive study, to include development of tools for change. In the meantime, it is hoped that this initial agenda-setting exercise will encourage further discussion and increased awareness of this critical issue in the European energy transition.

Acknowledgements The contributions of all the workshop attendees were invaluable in writing this piece. In addition to the authors, the attendees were Dr Christopher Decker, University of Oxford (Law and Economics); Dr Sylvaine Le Garrec, Consultant, Paris (Sociology); Dr Julie Goodhew, University of Oxford (Environmental Psychology); Professor Magdalena Habdas, University of Silesia (Law); Professor Miles Hewstone, University of Oxford (Social Psychology); Dr Sandra Passinhas, University of Coimbra (Law); Professor Vincent Sagaert, KU Leuven (Law); Professor Sergio Nasarre-Aznar, University Rovira i Virgili (Law); David Weatherall, Future Climate, London (Energy Policy); Dr Annemarie van Zeijl-Rozema, Maastricht University (representing the ACE Retrofitting project) (Sustainability Science/Co-creation).

Information on GREEAN-EU is available at http://futureclimate.org.uk/green/. We are also grateful to SHAPE ENERGY and the University of Oxford Law Faculty Research Support Fund for funding the Oxford workshop.

References

ACE Retrofitting. (2018). Accelerating Condominium Energy Retrofitting. [online]. Retrieved March, 2018, from http://www.nweurope.eu/projects/project-search/accelerating-condominium-energy-retrofitting-ace-retrofitting/

Ajzen, I. (1991). The Theory of Planned Behavior. *Organizational Behavior and Human Decision Processes, 50*(2), 179–211.

Akrich, M., Callon, M., & Latour, B. (Eds.). (2006). *Sociologie de la traduction: textes fondateurs*. Paris: Mines ParisTech, les Presses Sciences Sociales.

Bright, S., & Weatherall, D. (2017). Framing and Mapping the Governance Barriers to Energy Upgrades in Flats. *Journal of Environmental Law, 29*(2), 203–229.

Brisepierre G. (2011). *Les conditions sociales et organisationnelles du changement des pratiques de consommation d'énergie dans l'habitat collectif,* Thèse de sociologie sous la direction de Dominique Desjeux, soutenue le 19 septembre 2011 à L'université Paris Descartes.

Crozier, M., & Friedberg, E. (1977). *L'Acteur et le système.* Paris: Editions du Seuil.

Darley, J. M., & Latane, B. (1968). Bystander Intervention in Emergencies: Diffusion of Responsibility. *Journal of Personality and Social Psychology, 8*(4, Pt.1), 377–383.

Fox, E., Foulds, C., & Robison, R. (2017). *Energy & the Active Consumer—A Social Sciences and Humanities Cross-cutting Theme Report.* Cambridge: SHAPE ENERGY.

Golovtchenko, N. (1998). *Les copropriétés résidentielles entre règle juridique et régulation sociale. Contribution à une sociologie de l'action organisée,* Thèse de doctorat sous la direction d'A. Bourdin, université de Toulouse-II, pp. 464.

Hirst, E., & Brown, M. (1990). Closing the Efficiency Gap: Barriers to the Efficient Use of Energy. *Resources, Conservation and Recycling, 3*(4), 267–281.

Jaffe, A. B., & Stavins, R. N. (1994). The Energy-efficiency Gap: What Does It Mean? *Energy Policy, 22*(10), 804–810.

Janda, K. B., Wilson, C., Bartiaux, F., & Moezzi, M. (2015). Improving Efficiency in Buildings: Conventional and Alternative Approaches. In P. Ekins, M. Bradshaw, & J. Watson (Eds.), *Global Energy: Issues, Potentials and Policy Implications.* Oxford: Oxford University Press Ch. 9.

Le Garrec, S. (Eds.). (2014). *Qui sont les leaders énergétiques dans les copropriétés et quelles sont leurs stratégies, méthodes et bonnes pratiques pour favoriser la rénovation énergétique, Rapport final, Planète Copropriété, Plan Urbanisme Construction Architecture,* [online]. Retrieved March 2018, from, http://www.prebat.net/?Qui-sont-les-leaders-energetiques

Lefeuvre, M. P. (1999). *La copropriété en difficulté. Faillite d'une structure de confiance.* La Tour d'Aigues: Les Editions de l'Aube.

Low Energy Apartment Futures. (2016). *Improving the Energy Efficiency of Apartment Blocks, LEAF Final Report.* [online]. Retrieved March, 2018, from www.lowenergyapartments.eu

Lutzenhiser, L. (1993). Social and Behavioural Aspects of Energy Use. *Annual Review of Energy and the Environment, 18,* 247–289.

Matschoss, K., Heiskanen, E., Kranzl, L., & Atanasiu, B. (2013). Energy Renovations of EU Multifamily Buildings: Do Current Policies Target the Real Problems? In: Proceedings of the ECEEE 2013 Summer Study. Stockholm: Berg.

Raworth, K. (2017). *Doughnut Economics: Seven Ways to Think Like a 21st Century Economist.* London: Random House.

Sampson, R. J., Raudenbush, S. W., & Earls, F. (1997). Neighborhoods and Violent Crime: A Multilevel Study of Collective Efficacy. *Science, 277*(5328), 918–924.

Sniehotta, F. F., Presseau, J., & Araújo-Soares, V. (2014). Time to Retire the Theory of Planned Behaviour. *Health Psychology Review, 8*(1), 1–7.

Sorrell, S., Mallett, A., & Nye, S. (2011). *Barriers to Industrial Energy Efficiency: A Literature Review*. Vienna: United Nations Industrial Development Organization.

Tajfel, H., & Turner, J. C. (1986). The Social Identity Theory of Intergroup Behaviour. In S. Worchel & W. G. Austin (Eds.), *Psychology of Intergroup Relations* (pp. 7–24). Chicago, IL: Nelson-Hall.

Weatherall, D., McCarthy, F., & Bright, S. (2017). Property Law as a Barrier to Energy Upgrades in Multi-owned Properties: Insights from a Study of England and Scotland. *Energy Efficiency.* https://doi.org/10.1007/s12053-017-9540-5.

Open Access This chapter is licensed under the terms of the Creative Commons Attribution 4.0 International License (http://creativecommons.org/licenses/by/4.0/), which permits use, sharing, adaptation, distribution and reproduction in any medium or format, as long as you give appropriate credit to the original author(s) and the source, provide a link to the Creative Commons license and indicate if changes were made.

The images or other third party material in this chapter are included in the chapter's Creative Commons license, unless indicated otherwise in a credit line to the material. If material is not included in the chapter's Creative Commons license and your intended use is not permitted by statutory regulation or exceeds the permitted use, you will need to obtain permission directly from the copyright holder.

CHAPTER 7

Crossing Borders: Social Sciences and Humanities Perspectives on European Energy Systems Integration

Antti Silvast, Ronan Bolton, Vincent Lagendijk, and Kacper Szulecki

Abstract Our chapter brings together four Social Sciences and Humanities (SSH) scholars into a conversation about their research and policy engagements, working within History, Political Science, Sociology, and Science and Technology Studies. We develop a socio-technical perspective and turn that into a conceptual tool pack, to interrogate and explore the

---

A. Silvast (✉)
Department of Anthropology, Durham University, Durham, UK
e-mail: antti.e.silvast@durham.ac.uk

R. Bolton
School of Social and Political Science, University of Edinburgh, Edinburgh, UK
e-mail: Ronan.Bolton@ed.ac.uk

V. Lagendijk
Department of History, Maastricht University, Maastricht, Netherlands
e-mail: vincent.lagendijk@maastrichtuniversity.nl

K. Szulecki
Department of Political Science, University of Oslo, Oslo, Norway
e-mail: kacper.szulecki@stv.uio.no

© The Author(s) 2018
C. Foulds, R. Robison (eds.), *Advancing Energy Policy*,
https://doi.org/10.1007/978-3-319-99097-2_7

emerging concept of Energy Systems Integration (ESI) with a special interest in European energy integration. Our contributions include, first, advancing the concepts of socio-technical energy system and seamless web for our research topics. Second, we open up select frameworks for ESI using the socio-technical perspective and highlight very different interpretations of systems integration terminologies and their effects. Third, the chapter explores of how the production of scale matters greatly for integrated energy systems, from a variety of infrastructural scales to urban, national, and supranational scales. The chapter rounds up by suggesting ideas for future interdisciplinary research between SSH researchers and designers of more integrated energy systems.

Keywords Large technical systems • Infrastructures • Science and Technology Studies • Political Science • History

7.1 Introduction

What can academic disciplines from the Social Sciences and Humanities (SSH) offer to understand how Europe will reach its targets of affordable, reliable, and sustainable energy? Our chapter focuses on a variety of energy-related SSH research, and unpacks its implications for energy policy by bringing together four SSH scholars into a conversation about their research and policy engagements. We work within History, Political Science, Sociology, and Science and Technology Studies, and our projects have been funded by different councils and public as well as private donors in different national and international contexts: the EU Horizon 2020 and research councils and foundations in various countries (UK, Norway, Poland, and the Netherlands).

The key object of this chapter concerns the concept of Energy Systems Integration (ESI) with a special interest in European energy integration. This concept has been emerging in expert circles through specialised associations, conferences, and research projects for a number of years. The International Institute for Energy Systems Integration was founded in 2014 to address cross-sectoral integration of multiple energy systems, and mainly its technical challenges. It defines the concept as the following activity: 'Energy Systems Integration (ESI) is the process of coordinating the operation and planning of energy systems across multiple pathways and/or geographical scales to deliver reliable, cost effective energy services with minimal impact on the environment' (O'Malley et al. 2016, p. 1).

Systems integration visions also feature prominently in the European Commission's (2017) Strategic Energy Technology (SET) Plans—where the greater integration between heating, cooling, energy storage, energy efficiency, demand-side responses, and renewable energies appears as the next advancement in the context of 'smart' projects such as 'smart grids' or 'smart cities'. Similar visions have underpinned the EU's Energy Union (European Commission 2015), a strategy for European energy policy integration across cutting carbon to energy security, internal markets, innovation, and energy efficiency. The concept of ESI is also central to large research programmes including research universities and companies. One recent example is the collaboration of five UK universities and Siemens in the Centre for Energy Systems Intergration (CESI).

While ESI emerged mainly from technological fields, it involves significant SSH-related issues. In this chapter, we develop a socio-technical perspective to interrogate and explore a set of these issues. We are speaking primarily from our interest in Science and Technology Studies, while our varied disciplinary backgrounds represent a good sample of SSH. We argue that this subset offers well-developed concepts and tools for framing today's energy problems, and thus helps to formulate more thorough and robust solutions. This critical edge has policy relevance of its own. But in place of problem-solving blueprints, it helps to adjust the premises on which energy debates are set, the conceptual tool pack which we use to propose and design policies, and the perspectives for evaluating the pros and cons of different technical and political-administrative solutions.

Our chapter speaks to several themes from this edited book. The contributions include, first, addressing what SSH scholars understand by the concepts of *socio-technical energy systems* and their *technopolitics* (Hecht 2011) in matters of integration. These terms refer to an energy system that cuts across technological, political, social, disciplinary, jurisdictional, and organisational boundaries. This system forms what researchers name a *seamless web* (Hughes 1986), requiring strong collaboration between different academic disciplines and their knowledge practices in order to fully appreciate its technical and societal embeddedness (Winskel 2018). Second, we open up select frameworks for ESI using the socio-technical systems perspective and highlight the clear *interpretative flexibility* (Pinch and Bijker 1984) of various systems integration terminologies and their effects in so doing. Third, the chapter explores how the production of scale matters greatly for integrated energy systems in various meanings of the term—different infrastructural scales (Edwards 2003), the urban scale (Bolton and Foxon 2013), the national scale, and the supranational scale (Van der Vleuten 2004; Lagendijk 2008). The chapter rounds up by suggesting ideas for future interdisciplinary research and interactions between

SSH researchers and designers of more integrated energy systems. Each of these contributions is discussed in its own subsection in order.

The chapter was built upon an internal researcher workshop held at Durham University on 28 March 2018, between the four authors. Written up, this piece allows our readers to listen in on our conversations on our respective positions on Energy Systems Integration and engagement with academic SSH perspectives. Box 7.1 contains a more detailed explanation of the interdisciplinary workshop methodology for this chapter.

> **Box 7.1 Interdisciplinary workshop methodology for the development of research outputs**
>
> 1. All involved scholars identify relevant calls for workshop funding and develop an agenda to answer these concerns.
> 2. The workshop coordinator finds and contacts more scholars on the basis of the initial agenda.
> 3. The coordinator, together with other participants, develops a small number of orienting questions to the workshop discussion, building upon the initial agenda.
> 4. All scholars circulate a short abstract that addresses the questions, done at least one week before the workshop.
> 5. During the workshop, participants discuss the questions in order but also allow the discussion to move to new topics. The workshop starts with an introductory comment by each scholar and then allows all scholars to comment and discuss freely. The discussants are encouraged to present concrete problems and operational research concepts in order to avoid disciplinary biases shaping the discussion. Each participant takes notes, but the workshop is also recorded to ensure their accuracy.
> 6. After the discussion, the participants continue the workshop by jointly typing up key themes that emerged from the workshop.
> 7. After the workshop, the coordinator uses these notes to draft the first version of the paper.
> 8. Each participant comments on the draft before the submission.
> 9. The submission is subjected to peer review in the normal academic fashion.
> 10. Based on peer-review comments, the coordinator drafts the final output acknowledging comments from the other authors.

7.2 Integrated Socio-technical Systems and the Seamless Web Approach

Appreciating energy integration requires, at once, historical awareness of earlier integrated systems and the ways in which such integration has crossed boundaries from technical to different social arenas. With this issue in view, all SSH scholars involved in the workshop likened the electricity system to what historian Hughes (1986) has called a *seamless web*—an underlying idea shared in many wider Science and Technology Studies literatures (Bijker et al. 1987). At the core, the seamless web approach goes against professional or jurisdictional boundaries, disciplines, or categories of knowledge. Hughes argued that the integrating visions of inventors, managers, and engineers of technological systems always themselves crossed these dichotomies such as 'technology and science, pure and applied, internal and external, and technical and social' (Hughes 1986, p. 286). Furthermore, his research strongly emphasised that the system's components are commonly controlled and interact to fulfil a system goal. These goals can change over time—as can secondary policy goals and interpretations of energy issues—from security of supply to providing energy at the lowest possible price, or attaining sustainability aims.

What does this imply for how we should see energy systems? To us, it makes no sense to see the system as something purely technical and disconnected from other social and economic issues. Technological solutions only do not solve contemporary energy issues—think about local protests over energy grid extensions and wind farms or privacy risks some have associated with new energy smart metres. To focus on technologies only is to ignore the social embeddedness of technical artefacts and systems. Like Hughes (1986, p. 290) argued, we see the social and political embodied into the technical realm—something Gabrielle Hecht (2011) has labelled *technopolitics*. This means that a change in a system's goal will affect the components, and changes to components have consequences for the system's functioning.

When seeing them as seamless and socio-technical, systems immediately become interdisciplinary objects of study. Understanding historical shaping and current-day constellations of energy infrastructures requires technological, scientific, economic, sociological, anthropological, and political scientific lenses to fully grasp these systems. These concepts are very consequential for framing how we approach Energy Systems Integration. If we assume that integrators of energy systems are bounded by their disciplines, there is little need for more interdisciplinary research. But if we presume that the integrators of energy systems are problem-solvers to whom categories such as disciplines are soft and overlapping, then our analysis needs to be attuned to how that happens in practice, not simply assuming boundaries such as professions, disciplines, or nation-state borders. A socio-technical system like the grid combines the materiality of its physical infrastructures and the institutions and norms which emerge to govern it, as well as the broad set of practices that keep these together and enable the very functioning and use of the system. It is in the study of the establishment and change of these practices (and the related practical logic) that our contribution is anchored.

Hughes's (1986) historical work makes another important contribution in that visions for integrated systems, so common today, are not new. They go back to the invention of earliest electric power networks and other similar large-scale systems (Van der Vleuten 2004). Another important facet has been the *social* integration of engineers and inventors that aspired to build large energy systems. In Europe, the relevant engineering networks that envisioned a common electricity grid date to the early twentieth century (Lagendijk 2008). Hence, the ideal of supranational integration of energy systems predates the European Union (EU) by several decades. Energy integration does not simply equate to European integration along the lines of the EU.

As technology histories have shown, these earlier eras' energy systems builders were relatively easy to identify, being privileged actors (Van der Vleuten 2004) and varying from individual inventors, managers and engineers to investors and, especially with cross-border systems, international organisations of engineers. Concerning ESI, the relevant systems building is still emerging, and the actual systems integrators remain sometimes only implicit in discussions of the concept.

With regard to these systems builders, we seek to highlight three points. Firstly, at the moment, the concept of ESI seems to be emerging among policymakers and industrialists that might see themselves as systems integrators. Secondly, international and national networks and research projects that pursue Energy Systems Integration are very important for the social shaping of ESI. As such, they are often formed by academic researchers, especially modellers, analysts, and designers in engineering and scientific disciplines. But these communities can also include more policy-facing researchers and social scientists—especially when the integration of different disciplines (social, physical, and environmental) and interdisciplinary whole systems research are concerned (Winskel 2018). Thirdly, the commercial and economic principles of ESI remain underexplored but point to the importance of firms and energy market regulators as both shapers and potential beneficiaries of ESI (e.g. Farsi et al. 2008).

By bringing socio-technical infrastructures and these different designers and builders into a sharper focus, our perspective in this chapter joins and complements a particular approach of many of the texts in this edited book (Higginson et al. 2018, Chap. 5 in this collection; Genus et al. 2018, Chap. 9; Middlemiss et al. 2018, Chap. 2) and recent SSH discussions. This approach is the study of the meanings, competences, and materialities (e.g. household technologies and objects) underpinning social practices such as everyday energy use (Shove et al. 2012). The seamless web concept is also deeply interested in the everyday lives and practices of energy consumers but in a specific way. Its point is not to separate social practices and the energy system 'out there' but to view everyday life in the way in which problem-solvers of systems integration might see it. This means closely considering how the practices of energy users are anticipated and included in the design of more integrated system, whether it be in their technical specifications, inputs to whole energy systems models, or associated political values, standards, or legislations. Everyday life has a clear dynamic of its own that needs to be studied in its own ways and contexts; its integration and interaction with other energy system 'components' of various kinds are what concerns the seamless web approach in particular.

7.3 Frameworks for Energy Systems Integration

Like other contemporary energy terms—such as 'smart grids'—there is no single universal definition for Energy Systems Integration. Having said that, experts constantly associate specific values and valuations to more integrated energy systems, which are highly suitable for SSH interrogation as we show here. ESI, in its different guises, implies going beyond the limits of established organisational, jurisdictional, technological, and knowledge boundaries, which have framed centralised energy systems in Europe and other industrialised countries. This will be problematic for some actors, who may resist such changes; but it may also provide opportunities for other actors to reshape boundaries and create and capture the new sources of economic value which, according to the outputs of whole systems modelling, will arise from this integration.

In general, ESI is an emerging policy concept, with raising industrial interest, and with research being built around it as we speak. But it is also important to highlight that the concept has demonstrable *interpretative flexibility* (Pinch and Bijker 1984): different research initiatives may think of and interpret ESI differently, even offering very different designs for building these systems. It is not yet clear whether a coherent view on ESI will emerge or whether it will be constituted by many local and potentially incompatible interpretations. These interpretations can be imbued by very different social and political values: as SSH scholars know, energy technology is often designed in a way that enacts certain political goals. For example, in the European context, the values have ranged from grand visions of energy integration to pragmatic considerations of its technical and economic impacts and benefits (Bolton et al. 2018).

With these values and goals in view, the idea of ESI embeds the idea of a very inclusive energy system connecting everything. It also embeds the idea that incumbent infrastructures such as large gas and electricity distribution grids do not become redundant in the face of energy innovation. Rather, they can latch onto innovative energy technologies such as small-scale renewables and advances in 'smart grids' and other transitions that the systems are facing. Indeed, influential definitions very closely associate the value of ESI in minimising environmental impact (O'Malley et al. 2016) and hence energy transition. In so doing, systems that have already grown and become consolidated can find a new pathway by being more integrated with other emerging more sustainable energy systems. Finally, ESI seems to embody certain preference for system authorities and controllers—actors not as prominent for decades with the aspiration of liberalising infrastructure industries. More than market-based management, energy integration is an active process that actors partake in. Their role is coordinating this process including real-time operation and long-term planning. A major role will be played by analysers and designers of integrated systems that are meant to influence industries and policymakers, who we term *systems integrators* in this piece.

Very different system goals can be fulfilled by this ESI drawing upon various principles. One apparent goal follows an *engineering logic*. Here, ESI is done to optimise the inputs and outputs of infrastructure, where the isolation of energy carriers has led to resource inefficiencies, and better preplanning of these synergies would help realise the current energy transition. Another is a *market policy logic*. This logic is not directly about infrastructure but about how to organise the power market on a large scale so that it has energy resources and a large number of participants to enable competition and lowering prices. Both logics constitute a challenge to the *national energy sovereignty logic*, which drives many policy decisions in the EU and nationally (Szulecki and Westphal 2014; Szulecki and Kusznir 2018).

These logics are obviously different, but they can also overlap especially in the aspiration to optimise future energy systems. The term *techno-economic modelling* captures models of integrated systems that optimise energy systems for minimum costs and maximum welfare. One frequently used example of this is the TIMES model, which, as the Scottish Government (2016, p. 1) points out, is both 'a technical engineering approach and an economic approach'. This dual approach feeds into asking how much will the energy transition, for example, integrating an increasing amount of renewable energies in the power grid, cost to our society.

7.4 Scales of Energy Systems Integration

The values and logics of ESI are opened further by paying close attention to the variety of *scales* that they produce. All infrastructures embody multiple scales—power infrastructure, for example, exists at each household but also in utilities and transnational markets (Edwards 2003)—and one question for analysis becomes at which scale are we exploring the infrastructural system (Goldthau 2014). Another related concern, as scholars highlight in this edited collection, is that scales are not just fixed like administrative categories but produced by the very work of scaling via social processes, power, and contestation (Bridge et al. 2018, Chap. 11 in this collection).

With respect to ESI, there are many ways to do such scaling work. Integration of energy systems can happen *across national borders*, as well as across *regions*. The problematic of EU energy islands that lack international electricity or gas interconnectors to the EU's single energy system is one example of this scaling—where scale draws boundaries around regions and represents them as isolated from other scales. There is also integration *across energy carriers*—for example, hydrogen with gas—whether on a regional or supranational level such as the North Sea countries. Furthermore, energy integration can happen *across urban infrastructures*, on specific areas like a city, concerning how electrification comes together with heating or public transportation, for example. These scales are not fixed but change as the systems change; certain scales can be prioritised over others, depending on who controls the integration process.

On the other hand, at least as far as the EU and the power sector are concerned, a key reason for an apparent failure to achieve full ESI has been that processes of *techno-economic* and *political integration* have been happening at different speeds and, to some extent, at different scales, creating a disconnect. The former involves the building of the internal electricity market, with its infrastructural and procedural elements such as transmission lines, international interconnectors, market-coupling schemes, modelling of physical electricity flows, transfer of statistics, and common network codes. The elements of these processes are gradually put into place in the EU. The latter involves EU-level energy policy integration, which accelerated about a decade ago with the Third Energy Package (European Parliament and Council 2009) and led all the way to the Winter Package (European Commission 2016) following the Energy Union framework (European Commission 2015)—see, for example, Szulecki and others (2016). While obviously very closely related, it is testimony to the relevant problem-solving practices how often these two realities of integration stay apart. The recent Norwegian political debates and protests against ACER—the EU's official energy regulatory agency—offer an uncommon glimpse into a situation where techno-economic integration is actually associated with critical questions on centralised decision-making, power, and national sovereignty.

This leads to the final elephant in the room that needs to be addressed, especially when we speak of European Energy Systems Integration, the national scale. The centralised systems we have today were the result of choices made about scales, both *explicit* in political decisions and in business models and *implicit* in regulatory frameworks, industry codes, and engineering practices. From a techno-economic point of view, energy systems have very rarely respected national borders. The notion of national energy systems emerged only in the early 1910s, when many important cross-border micro-regional systems where already in place—and on the initiative of nationalistically minded governments (Schot and Lagendijk 2008). The engineers and managers of energy systems have sought for increased efficiency by integrating power resources across national borders, via synchronised power grids like Central Europe (Lagendijk 2008) and single power markets like the Scandinavian Nord Pool (Silvast 2017). From a more political point of view, however, states have almost always been tremendously important in matters of energy security of supply—that is, 'keeping the lights on'—and the exploitation of energy resources. The power sector was perhaps the last to be 'overtaken' by national governments, once borders between systems were drawn and utility monopolies emerged as the most popular business model for national energy provision (Bakke 2016).

The EU's Lisbon Treaty (Treaty of the Functioning of the European Union 2012) Article 194, which formally defined EU energy policy for the first time, even directly reflects this by making energy a *shared competence* between the EU institutions and the members (Szulecki and Westphal 2014). This principle means that the EU will act on energy policy in the areas of single markets, sustainability, and European security of supply. But other areas such as the use of national energy resources are decisions by the member states and can be fully legislated by them.

To sum up, we need to open up a discussion about the political dimensions of energy system scales. We need to interrogate at which scale(s) Energy Systems Integration is taking place and the implications of this. In so doing, SSH researchers need to be careful not to presume that the nation state will be the leading actor in matters of integration and to be alert as to how ESI might challenge the dominance of the national scale and create new cross-scale linkages.

7.5 Conclusion: Towards a Social Study of Energy Systems Integration

If ESI is to unfold as a research concept and a policy idea, there needs to be space made for a well-developed socio-technical perspective on its logics and practices. This implies reaching a number of related research goals. For one part, the socio-technical aspect of ESI is closely linked with developing a better understanding of energy consumption practices in integrated energy systems. The public reservations to infrastructure expansion, something which ESI often implies, offer an apparent example of how these issues can bite back if they are not sufficiently attended. SSH researchers can bring the 'seamless web' about by diagnosing and understanding these kinds of reservations. A considerable amount of work done under the banners of 'energy justice', 'energy democracy', and political participation (e.g. Jenkins et al. 2016; Szulecki 2018) has already made policymakers and engineers more aware that they themselves and energy technologies do not function in a socio-political vacuum.

At the same time, many benefits of ESI may actually not be achieved because of a mismatch between broader techno-economic dynamics and political and social processes. This situation calls for other tools and approaches by SSH researchers: a socio-technical perspective, which is more embedded within and engaged with the technical aspects of ESI research and practice. Rather than bracketing them out, this research should look at and problematise the socio-technical aspects of integrative energy systems and their interfaces to society, taking both historical continuities and discontinuities into account when studying these issues. Technical analysts themselves have provided detailed and useful definitions of ESI, but we argue their inclusive visions could be complemented with questions of socio-technical integration, the different interpretation of what this integration is for, and of the production of scale in ESI. Such new SSH language on ESI, we argue, is important for any researcher or policymaker interested in becoming a systems integrator or merely understanding the concept better, all over the world.

Acknowledgements We acknowledge the funding from the Scottish ClimateXChange project 'Scotland and the European Energy Union: A Socio-Technical Systems Perspective' and from the SHAPE ENERGY project, as well as the support of the Durham Energy Institute in co-organising the workshop this chapter builds upon.

References

Bakke, G. A. (2016). *The Grid. The Fraying Wires Between Americans and Our Energy Future*. New York: Bloomsbury USA.

Bijker, W. E., Hughes, T. P., & Pinch, T. (Eds.). (1987). *The Social Construction of Technological Systems: New Directions in the Sociology and History of Technology*. Massachusetts, CA: MIT Press.

Bolton, R., & Foxon, T. J. (2013). Urban Infrastructure Dynamics: Market Regulation and the Shaping of District Energy in UK Cities. *Environment and Planning A, 45*(9), 2194–2211.

Bolton, R., Lagendijk, V., & Silvast, A. (2018). Grand Visions and Pragmatic Integration: Exploring the Evolution of Europe's Electricity Regime. *Environmental Innovation and Societal Transitions*. https://doi.org/10.1016/j.eist.2018.04.001.

Edwards, P. N. (2003). Infrastructure and Modernity: Force, Time, and Social Organization in the History of Sociotechnical Systems. In T. Misa, P. Brey, & A. Feenberg (Eds.), *Modernity and Technology* (pp. 185–226). Cambridge, MA: MIT Press.

European Commission. (2015). A Framework Strategy for a Resilient Energy Union with a Forward-Looking Climate Change Policy. COM/2015/080 final.
European Commission. (2016). Clean Energy For All Europeans. COM(2016) 860 final.
European Commission. (2017). The Strategic Energy Technology (SET) Plan. Publication Prepared Jointly by European Commission's Directorates-General for Research and Innovation, Energy and the Joint Research Centre.
European Parliament and Council. (2009). Concerning Common Rules for the Internal Market in Electricity and Repealing Directive 2003/54/EC (Text with EEA relevance). Directive 2009/72/EC.
Farsi, M., Fetz, A., & Filippini, M. (2008). Economies of Scale and Scope in Multi-utilities. *The Energy Journal, 29*(4), 123–143.
Goldthau, A. (2014). Rethinking the Governance of Energy Infrastructure: Scale, Decentralization and Polycentrism. *Energy Research & Social Science, 1*, 134–140.
Hecht, G. (Ed.). (2011). *Entangled Geographies: Empire and Technopolitics in the Global Cold War*. Massachusetts, CA: MIT Press.
Hughes, T. P. (1986). The Seamless Web: Technology, Science, Etcetera, Etcetera. *Social Studies of Science, 16*(2), 281–292.
Jenkins, K., McCauley, D., Heffron, R., Stephan, H., & Rehner, R. (2016). Energy Justice: A Conceptual Review. *Energy Research & Social Science, 11*, 174–182.
Lagendijk, V. (2008). *Electrifying Europe: The Power of Europe in the Construction of Electricity Networks*. Amsterdam: Amsterdam University Press.
O'Malley, M., Kroposki, B., Hannegan, B., Madsen, H., Andersson, M., D'haeseleer, W., McGranaghan, M. F., Dent, C., Strbac, G., Baskaran, S., & Rinker, M. (2016). *Energy Systems Integration. Defining and Describing the Value Proposition (No. NREL/TP--5D00-66616)*. Golden, CO: National Renewable Energy Lab (NREL).
Pinch, T. J., & Bijker, W. E. (1984). The Social Construction of Facts and Artefacts: Or How the Sociology of Science and the Sociology of Technology Might Benefit Each Other. *Social Studies of Science, 14*(3), 399–441.
Schot, J., & Lagendijk, V. (2008). Technocratic Internationalism in the Interwar Years: Building Europe on Motorways and Electricity Networks. *Journal of Modern European History, 6*(2), 196–217.
Scottish Government. (2016). *A Scottish TIMES Model: An Overview*. [online]. Retrieved April 30, 2018, from http://www.gov.scot/Resource/0050/00508928.pdf
Shove, E., Pantzar, M., & Watson, M. (2012). *The Dynamics of Social Practice: Everyday Life and How It Changes*. London: Sage.

Silvast, A. (2017). *Making Electricity Resilient: Risk and Security in a Liberalized Infrastructure*. Abingdon and New York: Routledge.

Szulecki, K. (2018). Conceptualizing Energy Democracy. *Environmental Politics, 27*(1), 21–41.

Szulecki, K., & Kusznir, J. (2018). Energy Security and Energy Transition: Securitisation in the Electricity Sector. In K. Szulecki (Ed.), *Energy Security in Europe: Divergent Perceptions and Policy Challenges* (pp. 117–148). London: Palgrave Macmillan.

Szulecki, K., & Westphal, K. (2014). The Cardinal Sins of European Energy Policy: Nongovernance in an Uncertain Global Landscape. *Global Policy, 5*(1), 38–51.

Szulecki, K., Fischer, S., Gullberg, A. T., & Sartor, O. (2016). Shaping the 'Energy Union': Between National Positions and Governance Innovation in EU Energy and Climate Policy. *Climate Policy, 16*(5), 548–567.

Treaty of the Functioning of the European Union. (2012). Consolidated Version of the Treaty on the Functioning of the European Union, art. 194, C2012/326/01. http://filj.lawreviewnetwork.com/files/2011/10/EU_Citation_Manual_2010-2011_for_Website.pdf

Van der Vleuten, E. (2004). Infrastructures and Societal Change. A View from the Large Technical Systems Field. *Technology Analysis & Strategic Management, 16*(3), 395–414.

Winskel, M. (2018). The Pursuit of Interdisciplinary Whole Systems Energy Research: Insights from the UK Energy Research Centre. *Energy Research & Social Science, 37*, 74–84.

Open Access This chapter is licensed under the terms of the Creative Commons Attribution 4.0 International License (http://creativecommons.org/licenses/by/4.0/), which permits use, sharing, adaptation, distribution and reproduction in any medium or format, as long as you give appropriate credit to the original author(s) and the source, provide a link to the Creative Commons license and indicate if changes were made.

The images or other third party material in this chapter are included in the chapter's Creative Commons license, unless indicated otherwise in a credit line to the material. If material is not included in the chapter's Creative Commons license and your intended use is not permitted by statutory regulation or exceeds the permitted use, you will need to obtain permission directly from the copyright holder.

CHAPTER 8

A Complementary Understanding of Residential Energy Demand, Consumption and Services

Ralitsa Hiteva, Matthew Ives, Margot Weijnen, and Igor Nikolic

Abstract This chapter explores potential ways to implement, and benefits for policymaking of, the complementary use of two different types of modelling for analysing residential energy consumption and ethnographic research. The more traditional approach of techno-economic modelling is considered alongside agent-based modelling that incorporates both causal and intentional relationships; ethnographic approaches provide 'thick understanding' of the relationships between social and technical elements

---

R. Hiteva (✉)
Science Policy Research Unit, University of Sussex, Brighton, UK
e-mail: R.Hiteva@sussex.ac.uk

M. Ives
Environmental Change Institute, University of Oxford, Oxford, UK
e-mail: matthew.ives@smithschool.ox.ac.uk

M. Weijnen • I. Nikolic
Faculty of Technology, Policy and Management, Delft University of Technology, Delft, Netherlands
e-mail: M.P.C.Weijnen@TUDelft.NL; I.Nikolic@tudelft.nl

© The Author(s) 2018
C. Foulds, R. Robison (eds.), *Advancing Energy Policy*,
https://doi.org/10.1007/978-3-319-99097-2_8

and the environment. In doing so, the chapter builds on real examples from academic-policy engagement in the EU on energy demand, consumption and services. We examine three myths of the role of modelling in policymaking and propose practical ways of employing different types of modelling in a complementary way to increase policymakers' understanding of residential energy demand, consumption and services. Finally, we make three concrete recommendations for developing future interdisciplinary work on integrating social and technical models for informing policy.

Keywords Techno-economic models • Ethnographic research • Agent-based modelling • Policymaking • Understanding

'All models are wrong, some models are useful'
George Box 1979 (in Launer and Wilkinson 1979, p. 202)

8.1 Introduction

In our experience policymakers often use modelling (either themselves or by interacting with modellers) to help understand the potential impact of (in)action and identify useful points of intervention. By policymakers here we refer to civil servants who use or make models to inform government policy, while policymaking is considered as the organised attempt to select goals and methods for governmental action (Stevens 2011). However, models are often misunderstood and misused, in terms of what they can do and what models are suited to answer particular questions. The three most frequently encountered myths about the use of modelling in policymaking are outlined in Box 8.1. Greater understanding of models by policymakers is required as to what questions different models are good at answering and how they can best be used to inform energy policy.

Box 8.1 The mythology of modelling and policy
No single model or modelling process is best for policymaking. Instead, the process of designing a model is a decisive factor in what contribution the model has to policymaking (Kimbell 2011). In our experience the use of models and modelling in policymaking is often shaped by three myths.

Myth 1 is that *models produce objective evidence for policymaking*. Policymakers often expect that models will produce straightforward and concrete answers that can be used right away and that they will justify intervention choices. The culture of policymaking is dominated by the need to produce evidence that is statistically valid as opposed to 'policy by anecdote'. The use of evidence in policymaking is extensively criticised (van de Goor et al. 2017; Naughton 2005). What counts as evidence is itself a politically loaded discussion (Monaghan 2008). Often uncertainty in models and accompanying narratives is reduced to bullet points, diagrams, case studies, text boxes, infographics and 'killer charts' or removed entirely as a potential barrier to action. In the pursuit of certainty, policymakers often lose sight of emergent complexity and contradictions. Models providing inconclusive information (with multiple caveats, limitations and elaborated uncertainty) are seen as counterproductive to creating persuasive policy stories (Stevens 2011).

Myth 2 is that *models produce straightforward policy solutions*. That's why policymakers have a strong preference and expectation that models are not too complex to work with or understand, that model outputs are not too abstract and that models don't come with high levels of uncertainty (to be able to serve as evidence for recommended policy). Models are expected to simply 'speak for themselves' with their policy implications being immediately apparent. In fact, models do not provide answers that can be plugged into existing policy frameworks. Modelling is a socio-technical learning process (Bollinger et al. 2015), in which models and the insight they provide develop over time by designing and using them.

Myth 3 is that *policymakers need more data to take action*. Big data, large, software generated and machine-readable data sets, are preferred over 'thick data', smaller size but deeper data that might offer greater contextual insight produced through ethnographic research. Big data can really only be used with machine learning models, which are context- and knowledge-free and can only identify patterns in data without any understanding of how and why. Big data is often left to speak for itself: 'We don't need theory, we have data'. Investment in required software and hardware that can process big data can be substantial, in terms of cost, time and effort. Policy institutions can end up building, 'feeding' and investing in

> large models, which progressively offer less flexibility as they grow in size. Although civil servants display a high level of commitment to the use of evidence, they are rarely able to use the huge volume of evidence they are provided with (Stevens 2011). We argue that policymakers need more contextualised understanding rather than more big data. This is particularly important in the context of energy services where users' energy needs and wants are contextually embedded and thick understanding is needed to tell the difference between an eye twitch and a wink. A clear and well-worded case study (i.e. a narrative supported by evidence) can be just as effective in shaping policy as modelled outputs.

We use the concepts of energy demand, consumption and services to illustrate the different approaches and models that we need in order to 'see' demand, consumption and services. The discussion is focused on the use of techno-economic models and agent-based modelling (ABM) because they are commonplace tools used by policymakers in the UK, the Netherlands and Bulgaria, where most of our experience is based. The focus on ethnographic research has been identified as particularly welcome by the modelling community and increasingly needed for understanding energy services and their implications for policy.

This chapter is based on the authors' combined experiences in designing and applying different types of models for understanding energy systems, as an input in the policymaking process in EU member states. In addition, the authors received input from 12 critical friends based in UK and Dutch institutions, between February and April 2018, through holding discussions around specific questions and sharing experiences of involvement in policymaking through techno-economic modelling, ABM and ethnographic input and using these in research. All discussions were recorded and summarised.

In order to explore the role of models in policymaking, and their use, we will focus on the example of modelling energy demand, consumption and services. We distinguish between 'energy demand' as an economic abstraction, 'energy consumption' as an engineering abstraction and 'energy services' as an ethnographic abstraction. Other definitions of these terms exist, but here we treat *energy demand* as the amount of energy demanded of utilities, such as the energy demand from residential heating. This amount is only loosely connected with the actual behaviour of users, which can affect utilities' predictions of future demand. *Energy consumption* is what energy users actually consume, including the contributions of material factors such as house size and structure and user efforts to reduce energy use, such as through improved insulation. *Energy services* are what users actually want or need in terms of pure energy, such a 'heat', and incorporate behaviours such as users spending more time in places with shared heating (e.g. outside in the sun). Understanding the differences between these abstractions and the processes involved can enable a better understanding of the efficacy of energy efficiency improvements (GEA 2012), and the emergence of conservation movements or energy sufficiency (Herring 2009; Steinberger and Roberts 2010), and hence the design of more cost-effective energy supply regimes and better demand management programmes (Skea et al. 2011).

Residential energy use is an area undergoing significant changes in terms of policy interventions and practice, in the context of climate change, energy security, technical advancements and social and institutional dynamics. It is an important area for policymakers as it is a major contributor to overall electricity consumption and contributes significantly to peak demand, particularly during winter months in Europe (Ramírez Mendiola et al. 2017).

Policy development in the EU generally involves the collection of large volumes of data on user behaviour, for example, through electricity metering (Torriti 2014), to understand what motivates users' behaviour and what behaviours can and should be modified. Most energy policymakers are familiar with techno-economic models of energy markets that present energy demand as an aggregate function of the decisions of individual energy users, who are generally treated as fully informed and, if not fully rational, at least predictably irrational (Huntington 2011; Wilkerson et al. 2013)—assumptions that have been heavily criticised in economics, social theory and political analysis (Sawyer 2005).

The Energy Performance of Buildings Directive (2002/91/EC) requires member states to introduce energy certification in order to reduce energy consumption in buildings. Although the directive recognises measuring both the energy demand and real energy consumption, these two approaches can lead to substantially different values. Demand represents just a 'norm consumption' calculated from the physical characteristics of a building, while consumption depends on many different social, technical and environmental factors (Steixner et al. 2007). Informing policy decisions on models built to represent understanding of singular concepts like energy demand can leave certain contributing factors blank or hidden while shining a light on others. For instance, in the context of residential energy demand, an abnormally cold winter or non-standard building can create a vulnerable group of consumers who are hidden to policymakers because they are unable (for financial reasons) to increase their energy consumption to the level that their energy use provides the energy service of a comfortably heated home.

Environmental concerns led to the setting of an EU-wide target in the Renewable Energy Directive of 20% of all energy consumed to be provided by renewable sources by 2020. That will require unprecedented change in the energy sector. Demand side measures and behavioural changes can significantly reduce energy demand but require an understanding of user behaviour and the implications of the built and natural environment. Hence, a good understanding of energy demand is the cornerstone of the EU's future energy system (GEA 2012).

Looking at this problem from the perspective of *energy consumption*, and ultimately *energy services*, encourages an emphasis on this behaviour, looking at the energy that is desired 'for the services that it produces, such as space and water heating, cooling, lighting, cooking, etc.' (Hunt and Ryan 2015). In other words, 'useful energy' or 'useful work' of energy is being put to work in a way that is distinct from the energy use itself (Sorrell and Dimitropoulos 2008, p. 20). This ultimately is the behaviour that will be affected by any decisions made by policymakers and is therefore the behaviour that needs to be captured in the models used by policymakers.

8.2 What Model?

The ultimate goal of modelling is to gain insight and not necessarily the production of *a number* to satisfy policymakers. Models are a simplifying lens through which we look at the energy system. The choice of model we use defines what we see. Models help us focus on particular relationships between variables and how changes in different parts affect the entire system. Through a techno-economic lens, residences can be seen as subsystems of the national (even continental) energy system, which comprises a multitude of installations, both on the supply and demand sides, which are interconnected through pipelines and cables. Through a Social Sciences lens, the energy system is seen as a huge network of actors, including power and heat generators, network owners and operators, energy service providers, end-users, technology providers, energy authorities and policymakers. Through the ethnographic lens, we see residents acting in their home environment, following daily routines and interacting with appliances and installations in their built environment in ways that determine how and why they use energy. These lenses are complementary and together provide a richer, more sophisticated understanding of the energy system. Hereafter we will explore the models and understanding that have emerged from these three different lenses, and discuss their usefulness in policymaking.

8.2.1 Techno-economic Models

Techno-economic models are being increasingly relied on to better understand what combinations of measures, over what time frames and at what costs, will be required to meet energy policy goals (Winskel et al. 2011). Many policymakers and energy companies base their policies, tariffs and projects on data collected on user load profiles (Torriti 2014). Techno-economic models use such profile information as well as data on the underlying techno-economic characteristics of the residential energy systems. They model the underlying physical structure of the system, such as building sizes and types, and the ratings and engineered characteristics of appliances (Swan and Ugursal 2009). They focus on how much energy could be consumed by different types of households and what appliances are used and when. They provide insights into levels and timing of demand as well as long-term trends that can affect these.

Techno-economic models can be used to provide insights on residential energy demand and consumption for places, using data from a subset of the population. They specify causal relationships (based on the laws of nature) and engineering heuristics (e.g. scaling rules), and may use stochastic or exploratory analysis in dealing with uncertainties. They can come as optimisation or simulation models. Optimisation models search for the 'best' system configuration in a given normative scenario. Simulation models are used to develop forecasts of how the system may evolve under different scenarios, especially economic conditions. Many models informing energy policymakers are bottom-up optimisation models, based on detailed technical specifications of the subsystems and their components. Often these models have a legacy of use: policymakers are familiar with them and tend to trust their outputs (Ramírez Mendiola et al. 2017).

Techno-economic models are based on rigid mathematical formulations that can be solved analytically. In the current policy system, closed-form analytical solutions are automatically accepted as true rather than looking for alternatives that best describe what the *need* is, even if this problem is mathematically 'messy'. Furthermore, the level of aggregation models use cannot take individuals into consideration. In their focus on optimising the economic performance of the aggregate technical system, they generally assume homogeneous actors. As these models are based on past behaviour and lack the granularity needed to fully understand residential energy services, they are mostly incapable of anticipating or incorporating changes in user behaviour, new trends or technologies or system shocks.

Apart from numerical outputs, techno-economic models are also accompanied by narratives which explain the levels of uncertainty built in the model and the ways in which the model can be interpreted. While these supporting narratives provide boundaries within which the model makes sense, policymakers may not fully acknowledge these limitations in their decision-making. This is problematic, given the enormous effort invested in such large-scale energy system models (and their data sources), which creates a tendency for policymakers to stick with the established models and instead fit problem formulations to the capabilities of the available modelling platform.

8.2.2 Agent-Based Modelling (ABM)

For a deeper understanding of what drives the behaviour of energy systems, individual-level behaviours and relationships must be added to the picture of the system-level causal relationships that are captured in traditional techno-economic models. Actors in the energy system perform certain roles, defined by institutions (norms, conventions, legislation, regulation, market rules, etc.). They can pursue their own strategies within their limits (be they institutional, technological, capital, knowledge, information, etc.). Acknowledging that the continuous interactions between actors, and between actors and the physical system, shape the behaviour of the energy system implies a socio-technical perspective on the energy system. From this perspective, understanding the interactions between the social elements and between the social and technical elements and subsystems is indispensable for providing policymakers with an understanding of where and how to intervene in energy systems.

These individual behaviours can be translated into ABMs, in which actors are represented as heterogeneous software agents. Like real actors, software agents can be programmed to exhibit varying extents of bounded rationality, imperfect information access and risk aversion and be equipped with learning capabilities. ABM can thus be seen as a crossover between the Engineering and the Social Sciences, as it aims to simulate aspects of real actor behaviour, expressed as computer algorithms interacting with physical systems, in effect translating qualitative behaviour into quantitative data and processes.

In ABM, the technical components and subsystems can be modelled as they are with techno-economic models, and economic optimisation models can be included as parts of individual decision-making routines. ABM is eminently suited for bottom-up energy system simulations that explore emergent system behaviour, for example, under varying institutional regimes. For residents to be adequately modelled as agents in ABM, insights are needed into their behaviour as energy users and how this behaviour is shaped by their service needs and their specific socio-economic, cultural and physical environment. This is where ethnographic research can play an important role.

8.2.3 Ethnographic Approaches

Ethnography, as an approach to collecting data, entails a wide variety of methods. The artefacts, processes and relationships studied in ethnography will depend on the context of the study. Participant *observation* and *shadowing* can be thought of as 'traditional' methods of ethnography. While *observation* implies a level of detachment from what is being studied, shadowing could involve 'doing' in order to enhance understanding, as well as conducting interviews. These ethnographic approaches are apt for studying the relationship between human and non-human objects in the performance of everyday activities, and the meanings ascribed to various everyday activities. Unlike techno-economic models, these can investigate worldviews, sociocultural structures and the practices that shape behaviour, and help readers to immerse themselves into the world being studied. The purpose of ethnographic research is developing what Geertz (1973) calls 'thick understanding', 'a stratified hierarchy of meaningful structures' (p.6), which can help us tell a twitch from a wink, a fake wink or a parody.

This is helped by the multitude of forms of ethnographic data (some of which are far more emotive than numbers) which could include photos, videos, quotes, objects and diaries (Kimbell 2011). This characteristic of ethnographic approaches enables them to contextualise findings and analysis and to deliver situated understanding of particular issues. Ethnographic research is widely used in the academic and non-governmental realms, as a standalone and powerful means for studying and engaging with interdependent relationships occurring in everyday life. The usefulness of ethnographic research here is discussed only in the context of the contribution it can make for policymakers in understanding the limitations of models and working with them in the process of policymaking.

While ABM and techno-economic models are led by a set of rules and assumptions, ethnographic research can be hypothesis-free and exploratory, as well as more narrowly targeted. The residential focus on energy lends itself well to an analysis through ethnographic approaches, because it can incorporate different socio-technical drivers.

8.3 Bringing the Approaches Together

This section discusses how techno-economic and ABM models can be used together, along with ethnographic research, in a complementary way to enhance policy understanding of residential energy consumption, demand and services. Residential energy demand models have been classified into two main approaches, top-down and bottom-up (Swan and Ugursal 2009). The top-down approach makes use of historic sector-level time series data of energy consumption through an analysis of long-term trends in macroeconomic factors such as changes in GDP, employment, housing builds and climate. Bottom-up models on the other hand look to build up models of residential demand from a hierarchy of individual end-users, houses or groups of houses (Grunewald et al. 2016).

Techno-economic bottom-up models tend to provide a static representation of user behaviour and trends based on known drivers and hence are limited in their ability to expose new or unexpected behaviour or emergent trends in the system. A more dynamic understanding of user behaviour can be developed through the use of ABM and ethnographic shadowing and observation, as an additional class of bottom-up methods. Their applicability is well explained in the task of understanding the divide between energy demand, consumption and services, which is explained through the ability of each to incorporate behavioural or ethnographic information.

Through ethnographic input, informal (and otherwise hidden) activities and practices can be included in ABM (such as switching off the fridge in the winter to save on electricity). Observations and shadowing can add value to ABMs by informing the context of player's interactions and behaviour. In turn techno-economic models can provide physical and monetary processes and inform the description and powers of actors in ABMs.

Ethnographic outputs often do not fit the idea of evidence in policymaking due to potential subjectivity bias and lower number of cases studied. However, in the context of energy poverty, the abstraction and aggregation usually employed for ABM and techno-economic models create their own biases that can make it harder to 'see' the point in making policy to protect the few at the expense of many (see Middlemiss et al., Chap. 2, and Aberg et al., Chap. 4, in this collection). Middlemiss and others, for example, emphasise the need to include consideration of the 'lived experience' for understanding energy poverty. Ethnographic approaches can help here to create harder hitting and empathetic understanding of different aspects of energy policy.

Techno-economic models are already using ethnographic research to provide improved predictive strength, to validate system processes or to better understand conditions under which models may or may not apply (Swan and Ugursal 2009; Grunewald et al. 2016). However, it could be argued that ABMs are better equipped to utilise ethnographic research, providing behavioural insights captured in ABMs through bounded rationality in agents' behaviour, providing them with limited information and inconsistent preferences to add realism. Ghorbani et al. (2015) have adopted such a complementary approach to produce empirically grounded reasoning in ABM.

ABMs and ethnographic research are also best placed to provide a testing ground for proposed policies. They allow policies to be assessed in realistic situations, providing feedback on the efficacy of policies in differing environments, an understanding of any perverse or unforeseen outcomes or behaviours and insights into avenues for policy improvements. Ethnographic approaches can help policymakers understand how consumption changes with a new policy based in energy services, and ABM techniques can be used to scale up such behavioural responses to examine the implications of such emergent trends on the system as a whole.

Techno-economic models usually attempt to understand the more complex behaviour of individuals through increasing sample sizes in data collection. This is a limiting factor for ethnographic research due to the labour and costs involved in observation and shadowing activities. However, big data collection processes can be improved through ethnographic research to provide a better understanding of trends in the data (Strang 1996; Ladner 2012), including giving a reference baseline for detecting trends, informing the data collection processes (Wilson et al. 2015) and providing new or improved explanatory variables for analysis. Ethnographic approaches can develop thick data, which in turn can provide a contextually embedded understanding of the systems for ABMs. Equally, what someone might show or tell an interviewer might be very different to how they actually behave (Grunewald et al. 2016), and big data on actual appliance usage supplied by engineering and statistical studies can be complementary.

The value of incorporating ABMs into techno-economic energy demand modelling does not end with the improved access to ethnographic information. Techno-economic models and ABMs, both informed by ethnographic research, can thus be used in a complementary combination of quantitative and qualitative data. A complementary approach can help produce multiple perspectives and thus produce more nuanced understandings of the problems at hand, as well as possible solutions. Including ethnographic research on a par with techno-economic models and ABM can produce a richer type of data—that is, taking into account a wider range of social and environmental elements (Lockton et al. 2013)—and understanding.

The outputs produced by such a complementary approach will aid in gaining multidimensional insight that policymakers would not get otherwise, and that can open up opportunities for exploring alternative scenarios for policy intervention. It can lead to policymakers asking different questions of the models as well as contribute to the development of a multilayered narrative to accompany the results. Rather than welcoming more complexity as an output of different models and approaches, we encourage acknowledging difference and competing information and exploring such disjunctures further rather than keeping them hidden or discounting them completely.

8.4 Changing Attitudes and Practice Towards a Complementary Understanding: Recommendations to Policymakers

Increasingly policymakers need to understand the needs, capacities and perspectives of a variety of actors, including citizens, service users and beneficiaries, so that policies are fit for purpose and deliverable (Kimbell 2011). Therefore, energy policy needs a complementary understanding from techno-economic models, ABM and ethnographic observation and shadowing. However, this will require changing attitudes and cultures of policymakers and policymaking. Building on the three myths outlined in Box 8.1, the first three steps in this direction should involve:

1. *Appreciation of limitations*, rather than seeking evidence, as a modelling output. The policymaking process needs to develop greater awareness and appreciation of the limitations of different models available, both in terms of modelling process and outputs. This involves expanding policymakers' understanding of what constitutes evidence (to include a wider variety of data, especially thick data), where it comes from and how evidence is developed. The latter involves getting more directly engaged with the process of modelling and allowing for a complementary use of model types and approaches. This would imply a slower and more involved evidence-gathering process.
2. *Confronting modelling outputs* rather than looking for straightforward policy solutions. Policymakers need to appreciate the value of combining and 'confronting' outcomes of different modelling approaches, rather than seek non-controversial answers and evidence. This can be achieved through training in modelling, using models with researchers and modellers, as well as testing findings against ethnographic research. Confronting involves a more open process of evidence gathering for policymaking, as well as developing additional steps of confronting, reconciling and embracing controversy and complexity in the policymaking process.
3. *Building a community of interdisciplinary policymaking users* rather than focusing on providing more data for policymakers. Policymakers need to be enabled to work in interdisciplinary contexts and with interdisciplinary research teams. This involves a concerted effort and investment in building a community of policymakers willing to

accept and embrace interdisciplinary models and outputs. Within the UN and peacebuilding contexts, for example, managers are accountable for the data they use (or don't) to shape their decisions. This involves asking managers if they have the right kind information for the question at hand, enough information, and whether that information is reliable (Kimbell 2011).

The three recommended steps should be approached simultaneously, as they reinforce each other. For example, investing in building a community of interdisciplinary policymaking users will help foster the skills needed for confronting modelling outputs and appreciating the limitations of modelling and approaches used, and vice versa. However, further research is needed to fully understand how these three approaches can be brought together in a complementary way and what the limitations of this complementarity may be. To start with this research agenda can explore the extent to which complementary understanding can be achieved when strongly conflicting understandings are produced through different approaches (i.e. what are the limits of confronting and reconciling differential understandings within a policy area). Such research will need to be based on more interactions and experiments between policymakers, models and complementary approaches. These could take the form of open policymaking labs, and take more informal and direct formats, such as modelling and policymaking hackathons, studio workshops and walks.

The proposed approach can be considered at odds with dominant policymaking processes, where policymakers have to research, design and propose policy interventions within limited time frames, usually months. However, most policy is anticipatory rather than responsive, and considering unanticipated policy impact if policy is not fit for purpose, the case for 'slow policymaking' can go a long way in managing potential risks. Furthermore, with the development of a community of interdisciplinary policymaking users with experience in confronting modelling outputs, associated costs (in terms of invested time, effort and money) can decrease.

Acknowledgements This chapter was informed by personal experiences of the four authors in using different types of modelling and engaging with the policymaking process, as well as through discussions with the following critical friends: Kat Lovell, Brenda Boardman, Philipp Grunewald, Jose Ramirez-Mendiola, Sven

Eggimann, Pieter Bots, Mathijs de Weerdt, Neil Yorke-Smith, Milos Cvetkovic, Simon Tindemans, Kornelis Blok and Laurens de Vries. Reflections on ethnographic modelling used here were based on ethnographic work carried out as part of the Resilience and vulnerability at the urban Nexus of food, water, energy and the environment (ResNexus) project funded by a joint FAPESP-ESRC-NWO call, grant ref: ES/N011414/1. We are also very grateful for the thoughtful comments of the two reviewers and the editors.

REFERENCES

Bollinger, L. A., Nikolić, I., Davis, C. B., & Dijkema, G. P. J. (2015). Multimodel Ecologies: Cultivating Model Ecosystems in Industrial Ecology. *Journal of Industrial Ecology, 19*(2), 252–263.

Box, G. E. P. (1979). Robustness in the Strategy of Scientific Model Building. In R. L. Launer & G. N. Wilkinson (Eds.), *Robustness in Statistics* (pp. 201–236). New York: Academic Press.

van de Goor, I., Hämäläinen, R. M., Syed, A. M., Juel Lau, C., Sandu, P., Spitters, H. P. E. M., Eklund Karlsson, L., Dulf, D., Valente, A., Castellani, T., Aro, A. R., & on behalf of the REPOPA consortium. (2017). Determinants of Evidence Use in Public Health Policy Making: Results from a Study Across Six EU Countries. *Health Policy, 121*(3), 273–281.

GEA. (2012). *Global Energy Assessment—Towards a Sustainable Future*. Cambridge, UK: Cambridge University Press.

Geertz, C. (1973). *The Interpretation of Cultures. Essays by Clifford Geertz*. New York: Basic Books, Inc.

Ghorbani, A., Dijkema, G., & Schrauwen, N. (2015). Structuring Qualitative Data for Agent-Based Modelling. *Journal of Artificial Societies and Social Simulation, 18*(1).

Grunewald, P., Ramírez Mendiola, J. L., & Lane, K. (2016). Residential Demand Modelling—Time for Flexibility. Conference proceedings of *BEHAVE 2016, 4th European Conference on Behaviour and Energy Efficiency*. 8–9 September 2016, Coimbra, Portugal, pp. 1–15.

Herring, H. (2009). Sufficiency and the Rebound Effect. In H. Herring & S. Sorrell (Eds.), *Energy Efficiency and Sustainable Consumption: The Rebound Effect* (pp. 224–239). London: Palgrave Macmillan.

Hunt, L. C., & Ryan, D. L. (2015). Economic Modelling of Energy Services: Rectifying Misspecified Energy Demand Functions. *Energy Economics, 50*, 273–285.

Huntington, H. G. (2011). The Policy Implications of Energy-efficiency Cost Curves. *Energy Journal, 32*, 7–21.

Kimbell, L. (2011). Rethinking Design Thinking: Part 1. *Design and Culture*, *3*(3), 285–306.

Ladner, S. (2012). Ethnographic Temporality: Using Time-Based Data in Product Renewal. *Proceedings of EPIC 2012*, Washington, DC: American Anthropological Association.

Launer, R. L., & Wilkinson, G. N. (Eds.). (1979). *Robustness in Statistics*. New York: Academic Press.

Lockton, D., Bowden, F., Greene, C., Brass, C., & Gheerawo, R. (2013). People and Energy: A design-led Approach to Understanding Everyday Energy Use Behaviour. *Ethnographic Praxis in Industry Conference Proceedings*, *2013*(1), 348–362.

Monaghan, M. (2008). The Evidence-base in UK Drug Policy: The New Rules of Engagement. *Policy and Politics*, *36*(1), 145–150.

Naughton, M. (2005). Evidence-based Policy and the Government of the Criminal Justice System—Only If the Evidence Fits! *Critical Social Policy*, *25*, 47–69.

Ramírez Mendiola, J. L., Grünewald, P., & Eyre, N. (2017). The Diversity of Residential Electricity Demand—A Comparative Analysis of Metered and Simulated Data. *Energy and Buildings*, *151*, 121–131.

Sawyer, R. K. (2005). *Social Emergence: Societies as Complex Systems*. Cambridge: Cambridge University Press.

Skea, J., Ekins, P., & Winskel, M. (Eds.). (2011). *Energy 2050: Making the Transition to a Secure Low-Carbon Energy System*. London: Earthscan.

Sorrell, S., & Dimitropoulos, J. (2008). The Rebound Effect: Microeconomic Definitions, Limitations and Extensions. *Ecological Economics*, *65*(3), 636–649.

Steinberger, J. K., & Roberts, J. T. (2010). From Constraint to Sufficiency: The Decoupling of Energy and Carbon from Human Needs, 1975–2005. *Ecological Economics*, *70*(2), 425–433.

Steixner, D., Brunauer, W., & Lang, S. (2007). Demand vs Consumption—Analysing the Energy Certification for Buildings. *Journal of Building Appraisal*, *3*, 213–229.

Stevens, A. (2011). Telling Policy Stories: An Ethnographic Study of the Use of Evidence in Policymaking in the UK. *Journal of Social Policy*, *40*(2), 237–256.

Strang, V. (1996). Sustaining Tourism in Far North Queensland. In M. Price (Ed.), *People and Tourism in Fragile Environments* (pp. 51–67). London: John Wiley.

Swan, L. G., & Ugursal, V. I. (2009). Modeling of End-use Energy Consumption in the Residential Sector: A Review of Modeling Techniques. *Renewable and Sustainable Energy Reviews*, *13*(8), 1819–1835.

Torriti, J. (2014). A Review of Time Use Models of Residential Electricity Demand. *Renewable and Sustainable Energy Reviews*, *37*, 265–272.

Wilkerson, J. T., Cullenward, D., Davidian, D., & Weyant, J. P. (2013). End Use Technology Choice in the National Energy Modeling System (NEMS): An Analysis of the Residential and Commercial Building Sectors. *Energy Economics, 40*, 773–784.

Wilson, C., Stankovic, L., Stankovic, V., Liao, J., Coleman, M., Kane, T., Firth, S., & Hassan, T. (2015). Identifying the Time Profile of Everyday Activities in the Home Using Smart Meter Data. *ECEEE 2015 Summer Study Proceedings*, pp. 933–945.

Winskel, M., Anandarajah, G., Skea, J., & Jay, B. (2011). Accelerating the Development of Energy Supply Technologies: The Role of Research and Innovation. In J. Skea, P. Ekins, & M. Winskel (Eds.), *Energy 2050. Making the Transition to a Secure Low Carbon Energy System* (pp. 187–204). London: Earthscan.

Open Access This chapter is licensed under the terms of the Creative Commons Attribution 4.0 International License (http://creativecommons.org/licenses/by/4.0/), which permits use, sharing, adaptation, distribution and reproduction in any medium or format, as long as you give appropriate credit to the original author(s) and the source, provide a link to the Creative Commons license and indicate if changes were made.

The images or other third party material in this chapter are included in the chapter's Creative Commons license, unless indicated otherwise in a credit line to the material. If material is not included in the chapter's Creative Commons license and your intended use is not permitted by statutory regulation or exceeds the permitted use, you will need to obtain permission directly from the copyright holder.

PART III

Interplay with Energy Policymaking Environments

CHAPTER 9

Imaginaries and Practices: Learning from 'ENERGISE' About the Integration of Social Sciences with the EU Energy Union

Audley Genus, Frances Fahy, Gary Goggins, Marfuga Iskandarova, and Senja Laakso

Abstract This chapter aims (1) to identify problematic framings relating to the integration of Social Sciences and Humanities (SSH) research with the developing EU Energy Union and (2) to account for the practice of SSH-related energy policy integration with regard to the disciplines, actors, initiatives and processes involved. It articulates an imaginary of SSH and policy integration prevalent in Horizon 2020 funding calls relating to the

---

A. Genus (✉) • M. Iskandarova
Small Business Research Centre, Kingston University,
Kingston upon Thames, UK
e-mail: a.genus@kingston.ac.uk; m.iskandarova@kingston.ac.uk

F. Fahy • G. Goggins
School of Geography and Archaeology and Ryan Institute, National University of Ireland Galway, Galway, Ireland
e-mail: frances.fahy@nuigalway.ie; gary.goggins@nuigalway.ie

S. Laakso
Consumer Society Research Centre, University of Helsinki, Helsinki, Finland
e-mail: senja.laakso@helsinki.fi

© The Author(s) 2018
C. Foulds, R. Robison (eds.), *Advancing Energy Policy*,
https://doi.org/10.1007/978-3-319-99097-2_9

EU Energy Union, which prefigures what is asked of SSH. Implications of this imaginary for the framing, substance and process of energy policymaking and the role of SSH research therein are discussed. An alternative imaginary is depicted, based on reflection on 'European Network for Research, Good Practice and Innovation for Sustainable Energy' (ENERGISE), a three-year, pan-European Horizon 2020-funded project being undertaken by the authors and other partners. The conclusion identifies priorities which need to be addressed in future Horizon 2020-funded research, centring on further probing of alternative imaginaries of, and approaches to, eliciting energy policy integration of SSH.

Keywords ENERGISE project • Energy research • Horizon 2020 • Imaginaries • Policy integration • Social sciences

9.1 Introduction

The European Commission is concerned to realise the potential contribution of Social Sciences and Humanities (SSH) to the achievement of objectives across a range of societal challenges, for example, by establishing the integration of SSH as a cross-cutting theme across funding programmes such as Horizon 2020. However, within EU energy research and policymaking, SSH remain to be effectively integrated (Foulds and Christensen 2016). SSH has suffered in comparison with STEM disciplines in energy research funding and perceptions of policy relevance.

In the academic literature, it has been noted that social 'dimensions' of energy are frequently neglected while there is greater emphasis on material and technical questions, something Sovacool et al. (2015) refer to as 'disciplinary chauvinism'. Moreover, SSH research is eclectic, including that which could inform energy research and policy at EU and national levels. The SHAPE ENERGY platform lists 20 SSH disciplines, including both Business and Theology. Undervaluing this variety may lead to neglect of core aspects of the climate change/energy challenge, such as moral questions about human needs, or overemphasis of technical, material and narrowly behavioural aspects (Castree 2016; Shove 2014).

The lack of integration highlighted above is due partly to the nature of imaginaries of energy-SSH adopted by policymakers and funders. The chapter argues that the primary reason for this shortcoming concerns the 'imaginary' of SSH energy policy integration that has been institutionalised in EU funding calls and prefigures the aims, roles and approaches to be adopted in funded projects, as well as their expected impact on policy. The article suggests that an alternative imaginary is possible and compares prevailing and 'new' contending, though interdependent, imaginaries.

The chapter is organised as follows: Sect. 9.2 discusses what is meant by 'imaginaries' and 'integration' in relation to SSH research. Section 9.3 identifies the imaginary of SSH energy policy integration manifest in selected Horizon 2020 work programmes. Section 9.4 considers the imaginaries of SSH integration implicated with proposing and executing the European Network for Research, Good Practice and Innovation for Sustainable Energy (ENERGISE) project,[1] a large-scale, three-year (2016–19) project funded under the European Commission Horizon 2020 framework programme. Finally, Sect. 9.5 reflects on what may be learned from the above regarding the need for, and institutionalisation of, new imaginaries capable of enhancing the integration of 'softer' SSH approaches in research and policy. Such imaginaries, research and policy would recognise the importance of citizen action, and energy-related cultures and practices, to the transformation of unsustainable lifestyles across the EU.

9.2 Understanding Imaginaries, Integration and SSH Research

9.2.1 Imaginaries

A growing literature has developed around 'sociotechnical imaginaries'. Sociotechnical imaginaries are defined as 'collectively held, institutionally stabilized, and publicly performed visions of desirable futures, animated by shared understandings of forms of social life and social order attainable through, and supportive of, advances in science and technology' (Jasanoff 2015, p. 4; c.f. Castoriadis, 1987). '[T]hey reside in the reservoir of norms and discourses, metaphors and cultural meanings out of which actors build their policy preferences' (Jasanoff and Kim 2009, p. 123) and in 'project visions of what is good and worth attaining' (Sovacool and Hess 2017, p. 719).

Jasanoff and Kim (2009) refer to six dimensions that may be employed in the analysis of sociotechnical imaginaries, which are adapted to inform the work of this chapter. The dimensions are (1) the framing of societal challenges and opportunities which SSH energy research might address, (2) policy focus (e.g. as present in the text of calls for funding), (3) controversies (over what do they arise?), (4) stakes (what could be won or lost in resolving controversies?), (5) closures (how the issues at stake are or will be resolved) and (6) civic epistemologies (e.g. the prominence and legitimacy of quantitative and qualitative research methods and processes governing relations among state authorities, experts and civil society).

Methodologically, sociotechnical imaginary approaches are well suited to critical investigation of the meanings attached to, institutionalisation of and change in EU research funding priorities and policies. Drawing on the analytical framework presented above, this chapter represents a novel application of the sociotechnical imaginary approach to the analysis of EU energy and research funding policies and integration of SSH research.

9.2.2 'Integration'

In energy-related research, SSH integration with policy is often addressed as part of wider debates about energy transitions, sociotechnical systems design and the role of SSH in interdisciplinary research (Rochlin 2014; Cooper 2017; Castree and Waitt 2017; Stern 2017). There may be differences between qualitative SSH researchers and policymakers regarding what qualitative SSH can realistically achieve and over what timescales (Rochlin 2014; Castree and Waitt 2017). For example, the current conceptualisation of the idea of 'policy impact' reflects a rather narrow understanding of the role and integration of SSH research, one which is oriented towards specific societal problems defined in instrumental terms set by policymakers rather than collectively determined through inclusive deliberation among a range of stakeholders. This approach often leads to qualitative SSH being treated as secondary to natural science but also to the 'harder', more 'scientific' of the SSH disciplines such as Economics. This approach may also neglect the wider impact that SSH has in influencing policy agenda and governance (for politics rather than policy) (Castree and Waitt 2017).

A broader notion of integration implies inclusion of different disciplinary perspectives in research policy and funding (e.g. Horizon 2020). Here, integration of SSH is commonly viewed as integration with STEM in interdisciplinary programmes and projects, which poses certain challenges as hierarchies and asymmetries still persist (Pedersen 2016). Pedersen (2016) illustrates this point with a critical analysis of the Horizon 2020 programme, suggesting that the political rhetoric of interdisciplinarity is driven by user needs and political incentives rather than bottom-up research interests. Furthermore, interdisciplinarity is not a magic bullet solution (Fox et al. 2017); even between SSH disciplines, insurmountable disagreement often exists. Hence, such integration of SSH approaches may be impracticable and/or ineffective.

The argument here is for an imaginary in which EU energy policy integrates qualitative SSH which recognises the collective nature of social practice and its implication for building energy policies and governance on a renewed understanding of energy demand and how it may be reduced. There is some way to go before such an imaginary may be said to predominate, as the next section on EU work programmes and funding calls will verify.

9.3 Imaginaries and SSH Integration: Analysing EU Energy Research Funding Calls

Energy-SSH disciplines have been underutilised by policymakers, in the European context and beyond, in spite of their considerable potential.

9.3.1 *Integration of SSH: The Text of Three Horizon 2020 Work Programmes*

In the text of the 2014–15 Horizon 2020 work programme for Secure, Clean and Efficient Energy (hereafter 'SC3'),[2] 'social sciences' is mentioned once. This is in connection with a specific challenge requiring socioeconomic research on energy efficiency (EE 12–2014), wherein (on p. 25) energy efficiency is stated to be 'playing a growing role in local, national and European policy development. It is a complex issue spanning different disciplines including engineering and social sciences'.

In addition, there is a reference to the need for applicants to 'take gender issues into account as well as existing macroeconomic and microeconomic models and results of socio-economic sciences and humanities' (again in EE 12–2014: socioeconomic research on energy efficiency, on p.25), with 'a specific priority [being] given to the development of microeconomic analysis of the latest energy efficiency measures'.

Note in the above the slippage in language across the few mentions of 'social sciences', 'socio-economic sciences' and 'humanities'. Later work programmes more consistently refer to 'SSH', possibly eliding differences between at least 20 different disciplines and arguably employing a formulation which tags humanities on to social sciences.

In 2016–17, SSH became more prominent. There were two references to SSH in headings in the competitive and low carbon energy call within the SC3 work programme[3] and a stronger and more frequent appeal to SSH both in the introductory 'blurb' of the programme and in the subsequent text. For example (on p. 10), it is considered that 'New approaches will therefore have to be stimulated as regards business models, competitive services, and an increasingly smart and dynamic system utilizing, wherever possible, a multidisciplinary approach, integrating different Social Sciences and Humanities fields'.

Reference is also made (on p.106) to the need for 'solid involvement of Social Sciences and Humanities and local communities and civil society to understand best practices and to increase knowledge'. Further, it is recognised (p.126, in relation to a European platform for energy SSH)[4] that 'researchers in the Social Sciences and Humanities (SSH) have a particular expertise in analysing and understanding deep change and in designing innovation processes, including social innovations' and that 'they *must* [our italics] play a stronger role in addressing energy-related challenges. Accordingly, SSH aspects *must* be better integrated into all stages of the research process'. However, other references to SSH continue to exemplify weaker integration of SSH, in ways which do not depart significantly from the 2014–15 work programme.

In the text of the 2018–20 SC3 work programme,[5] there is a continuation of the stronger version of SSH integration discussed above. Indeed, there is a prescriptive tone used throughout the text in relation to SSH. In a number of cases, it is stated that funded projects 'will use' or make 'paramount' use of techniques and methods of SSH to identify relevant stakeholders and analyse needs and increase awareness and assess impact on society.[6] At the same time, there are appeals to 'balance', 'i.e. [p]roposals will combine the relevant scientific and technological elements of these fields with relevant Social Sciences and Humanities'.[7] There remains a sense of SSH being necessary yet subordinate to science and engineering, however, as in previous work programmes. For example, the text outlining LC-SC3-RES-28-2018-2019-2020: Market Uptake support states that the 'complexity of [the] challenges… calls for multidisciplinary research designs, which should include contributions *also* from the social sciences and humanities' (pp. 71–72, our italics).

9.3.2 Imaginary of SSH in Horizon 2020 SC3

This section discusses the imaginary of SSH in SC3 work programmes, categorised on the basis of the six dimensions of sociotechnical imaginaries by Jasanoff and Kim (2009): framing of risks and opportunities, policy focus, controversies, stakes, closures and civic epistemologies. Although there is now a greater appreciation of the contribution of SSH disciplines and approaches, overall there remains a tendency to frame EU energy challenges and research as primarily technical in character. Further, the contribution of SSH is typically framed in relation to risks concerning the need for social acceptability or resistance to change.

The dominant policy focus is the growing role of energy efficiency in EU policy development and market uptake of renewable energy technologies. These are posed in the context of controversies or challenges relating to ensuring behavioural change and improved consumer choices, for example, achieved through the implementation of 'ICT-based solutions' in a problem-solving model (p.28, Horizon 2020 2016–17 SC3 work programme, call EE-07-2016-2017: Behavioural change toward energy efficiency through ICT).

At stake are the achievement of EU climate change targets, the competitiveness of the EU within the global renewable energy sector and, increasingly, how to ensure the buy-in of citizens/consumers across the EU within processes of responsible innovation, which has become a working principle underpinning EU research and innovation. Closures are framed in terms of contributions that funded research can make to EU or national energy policy development, predicated either on changing practice cultures in a participatory manner or nudging individual consumers to make 'better' choices.

Finally, in relation to civic epistemologies, there is an emphasis on the production of knowledge capable of shedding light on factors enabling individual consumers or households to make better energy choices. Such knowledge may involve or require the particular expertise of social scientists, working with local communities.

9.4 IMAGINARIES AND INTEGRATION: THE CASE OF ENERGISE

9.4.1 Introducing ENERGISE

ENERGISE is a three-year research project funded by the European Commission under the Horizon 2020 programme within the SC3 societal challenge, which aspires to strengthen the integration of SSH with the emerging EU Energy Union. ENERGISE aims to achieve a greater understanding of the social and cultural influences on residential energy use in Europe and to develop and test novel bottom-up approaches for reducing household energy demand across different contexts.

While the project is interdisciplinary in nature, incorporating various academic approaches that focus on a common goal, it is also transdisciplinary, insofar as it incorporates nonacademic and experiential knowledge in the research process (Holbrook 2013). The project incorporates perspectives from various stakeholders including businesses, NGOs, policymakers, government agencies and community groups, all of which are represented on the project's advisory panel. Project partners have also liaised with numerous local and national groups, from national energy agencies to local authorities and interest groups, contributing to the co-creation of knowledge. The inclusion of diverse perspectives increases the likelihood that the project outputs will be applicable and relevant for a wider audience and in various contexts and facilitates the exchange of knowledge between scientists, policymakers, practitioners and civil society. Meeting the needs of different audiences that may have very different requirements, as well as competing perspectives, presents a number of challenges. For example, it requires the production of a range of tailored outputs (Rau et al. 2018). Open communication and feedback between project partners and regular two-way engagement with external stakeholders are considered key to overcoming these challenges.

The project adopts an experimental Living Lab approach, which aims to generate knowledge in a 'real-world' setting that addresses the complex problem of excess energy use (Heiskanen et al. 2018). The nature of this kind of research setting is open-ended and allows for some degree of creative flexibility regarding design/implementation by not having at the outset a particular defined template for ENERGISE Living Labs (ELLs) or a fixed image of what 'community' or co-creation entail. While this also requires intense coordination and debate among the project partners, the flexibility enables the production of a contextually and culturally sensitive ELL design which could stand a better chance of being more successfully implemented and hence make a greater contribution to broader sustainability transformation.

9.4.2 Comparing Imaginaries of SSH

Table 9.1 summarises and compares imaginaries between the ENERGISE project proposal and the 2014–15 Horizon 2020 SC3 work programme. The comparison is elaborated in the following paragraphs.

Framing of societal challenges/risks and opportunities: The ENERGISE project is broadly framed as a response to perceived failures of technological approaches to address the problem of excessive residential energy use and related CO_2 emissions: despite increases in energy efficiency, the total energy use in households continues to grow. The main societal challenge in ENERGISE is the need for a sustainable and responsible energy transition rather than social acceptability of energy-efficient technologies.

Policy focus: ENERGISE aims at improving decision-making and providing recommendations for national and EU-level policy that derive from better understanding of socially shared practices rather than a concern to 'nudge' choices and diffuse low carbon or renewable energy technologies.

Table 9.1 Comparing imaginaries: ENERGISE project proposal and H2020 SC3 (2014–15)

	ENERGISE proposal	Horizon 2020 SC3 2014–15
Framing risks	Technological failure; need to understand energy-related practice cultures	Technical challenges; need for social acceptability
Policy focus	Changes in energy practice cultures; participatory governance	Energy efficiency; increase uptake of renewable energy technologies
Controversies	Competing understandings of (how to change) energy-related practice cultures	Top-down approach to ensuring behaviour change; consumer choice
Stakes	Realising the energy transition through responsible governance	Competitiveness; buy-in of customers
Closures	EU and national policies and interventions predicated on changing practice cultures	Technical energy efficiency innovations; policy measures to 'nudge' individual choices
Civic epistemologies	Understanding energy practice cultures through co-creation of knowledge	Enabling consumers to make 'better' energy choices

Source: authors' own application of the framework proposed by Jasanoff and Kim (2009)

Policy integration in the project is present both locally by collaboration with and empowerment of local actors in ELLs and co-creation of contextually relevant knowledge and nationally and cross-nationally via the sharing of knowledge of practice cultures and cross-cultural good practices for researching and transforming energy use. SSH research provides means to investigate and analyse both individual- and collective-level differences within and across national sites, the effectiveness of Living Lab approaches, and energy-related practices.

Controversies recognised in the project proposal arise over bottom-up and prevalent top-down approaches to energy demand reduction. The first controversy is related to the localised and contextualised aspects of energy use and diverse practice cultures and need to focus on them on the one hand and the need for comparable outcomes across Europe on the other. The second controversy considers the focus on co-inquiry (Genus 2014) and co-creation processes (and multiway engagement) with local stakeholders versus (inter)national energy governance. Understanding social norms related to energy use requires in-depth and qualitative approaches. Shifting these collective norms cannot be done within one research project, but attention needs to be paid to ways to upscale the research findings.

Stakes: In contrast to a concern about consumer buy-in and EU competitiveness, the imaginary epitomised by ENERGISE contributes to co-creation of knowledge about energy demand reduction and sufficiency of energy use. It engages with issues of democratisation and empowerment in a responsible approach to energy governance. At the same time, the novelty of the approach, ambitious goals and cross-national comparisons might lead to the need for simplifications and compromises in the research process.

Closures: The ENERGISE project emphasises that it goes beyond what is typically asked for in Horizon 2020 energy work programmes, which are predicated upon the quest for greater energy efficiency, economic analyses and technical innovation. Based on developing knowledge of energy-related practice cultures, ENERGISE hopes to influence the setting of future policy agendas for social inquiry and shape future research, as well as to contribute to improved decision-making at different policy levels and the development of Energy Union.

Civic epistemologies: ENERGISE aims to improve the qualitative understanding of different energy-related practice cultures, as well as the differences between individual and collective behaviour and data informing knowledge of factors influencing differences between these foci.

Qualitative methods and the Living Lab approach are used to, for example, reveal underlying dynamics such as qualitative changes in energy demand or shifts in daily routines due to ruptures and change initiatives. The cross-disciplinary and co-creative approach to working with the knowledge of energy citizens allows for going beyond conventional, state-of-the-art research and policy with its emphasis on providing consumers with better information on which to base decisions.

9.5 Conclusion: Towards a New Imaginary of SSH Energy Research

This chapter was written out of a concern that qualitative SSH was not being sufficiently or effectively integrated into EU energy research and policy. The chapter argues that this shortcoming is connected with the playing out of a certain imaginary of energy research and its integration with policy. Such an imaginary infuses programmatic calls for funding under Horizon 2020. These are also to be seen in the writing of Horizon 2020 project proposals such as that for ENERGISE. However, both the proposed and implemented designs of a project such as ENERGISE demonstrate the potential of a new imaginary for the integration of 'softer' SSH with energy research and policy in and within the EU. Hence 'new' and prevailing institutionalised imaginaries are at the same time interdependent and compete with each other. Thus, as proposed, ENERGISE reproduces the established imaginary—in attempting to gain high scores from project proposal evaluators for relevance to the aims of a funding call—even as its researchers propose a contending view. This tension continues into the conduct of the project, which still needs to satisfy programme aims while making the case for a new imaginary.

In terms of learning from ENERGISE about the nature of future European energy work programmes and funding calls that may require SSH research, it is important to note that this chapter is not advocating that the ENERGISE project should serve as a template for others. Fundamentally, the trajectory of any research project is contingent upon a range of project internal and external factors (Rau et al. 2018). While other studies have called for Horizon 2020 programmes more generally to embrace SSH (e.g. Bitterberg 2014), the case of ENERGISE serves to highlight some core issues specifically regarding the effective integration of qualitative SSH energy research with the developing EU Energy Union.

The ENERGISE proposal was developed in response to a Horizon 2020 SC3 funding call which was premised on a problem-solving model, at the centre of which lay concerns about how to effect behaviour change on the part of energy users and how to promote innovation of renewable energy or energy efficiency technologies. Such an imaginary may well be imbued with strategic intent, in which SSH can contribute by shedding light on behavioural aspects of energy or energy technologies, for example, in relation to economic inducements or interventions required to galvanise greener consumer preferences. However, it has not been as successful as hoped at addressing the need to understand in greater depth the antecedent conditions of consumption, which may be implicated with energy-related practice cultures. To the extent that this is so, EU SSH energy research funding has been calling upon a limited part of the repertoire of SSH, which if better and more fully utilised could enhance EU energy policymaking. In the forthcoming European Commission Horizon Europe (ninth) framework programme, which will be launched in January 2021, this could be addressed by adapting the language of energy work programmes and funding calls in favour of under-represented aspects of SSH. This could be achieved by prioritising more fulsome interdisciplinarity or transdisciplinarity, flexibility in research design and co-creation of knowledge in experimental sites (such as Living Labs), capable of revealing, understanding and transforming diverse energy-related practice cultures.

Acknowledgements In preparing this chapter, the authors have drawn heavily on work conducted for the European Network for Research, Good Practice and Innovation for Sustainable Energy (ENERGISE) project, which receives funding from the European Union's Horizon 2020 Research and Innovation programme under Grant Agreement No 727642.

Notes

1. The five co-authors of the chapter are active researchers on the ENERGISE project, with backgrounds in different SSH disciplines: Innovation, Human Geography, Science and Technology Studies, Sociology and Environmental Studies. See www.energise-project.eu for more details.
2. See 2014–15 energy work programme, which is currently accessible here: https://ec.europa.eu/research/participants/data/ref/h2020/wp/2014_2015/main/h2020-wp1415-energy_en.pdf.

3. See 2016–17 energy work programme, which is currently accessible here: http://ec.europa.eu/research/participants/data/ref/h2020/wp/2016_2017/main/h2020-wp1617-intro_en.pdf.
4. See call text for LCE-32-2016: European Platform for energy-related Social Sciences and Humanities research.
5. See 2018–20 energy work programme, which is currently accessible here: http://ec.europa.eu/research/participants/data/ref/h2020/wp/2018-2020/main/h2020-wp1820-energy_en.pdf.
6. See call text for LC-SC3-NZE-3-2018: Strategic planning for CCUS development.
7. See call text for LC-SC3-CC-5-2018: Research, innovation and educational capacities for energy transition.

REFERENCES

Bitterberg, C. (2014). Report on the Integration of Socio-economic Sciences and Humanities (SSH) in Horizon 2020 Deliverable 3.3 of the net4society Project Funded Under the *European Union's Horizon 2020 research and innovation programme* GA No: 320325. http://www.net4society.eu/

Castoriadis, C. (1987). *The Imaginary Institution of Society*. Cambridge, MA: MIT Press.

Castree, N. (2016). Broaden Research on the Human Dimensions of Climate Change. *Nature Climate Change, 6,* 731.

Castree, N., & Waitt, G. (2017). What Kind of Socio-technical Research for What Sort of Influence on Energy Policy? *Energy Research & Social Science, 26,* 87–90.

Cooper, A. C. G. (2017). Building Physics into the Social: Enhancing the Policy Impact of Energy Studies and Energy Social Science Research. *Energy Research & Social Science, 26,* 80–86.

Foulds, C., & Christensen, T. H. (2016). Funding Pathways to a Low-carbon Transition. *Nature Energy, 1*(7), 1–4.

Fox, E., Foulds, C., & Robison, R. (2017). *Energy & the Active Consumer—A Social Sciences and Humanities Cross-cutting Theme Report*. Cambridge: SHAPE ENERGY.

Genus, A. (2014). 'Coinquiry' for Environmental Sustainability: A Review of the UK Beacons for Public Engagement. *Environment and Planning C: Government and Policy, 32,* 491–508.

Heiskanen, E., Laakso, S., Matschoss, K., Backhaus, J., Goggins, G., & Vadovics, E. (2018). Designing Real-world Laboratories for the Reduction of Residential Energy Use: Articulating Theories of Change. *Gaia, 27*(1), 60–67.

Holbrook, J. B. (2013). What Is Interdisciplinary Communication? Reflections on the Very Idea of Disciplinary Integration. *Synthese, 190,* 1865–1879.

Jasanoff, S. (2015). Future Imperfect: Science, Technology, and the Imaginations of Modernity. In S. Jasanoff & S.-H. Kim (Eds.), *Dreamscapes of Modernity: Sociotechnical Imaginaries and the Fabrication of Power* (pp. 1–33). Chicago, IL: University of Chicago Press.

Jasanoff, S., & Kim, S.-H. (2009). Containing the Atom: Sociotechnical Imaginaries and Nuclear Power in the United States and South Korea. *Minerva, 47*, 119–146.

Pedersen, D. B. (2016). Integrating Social Sciences and Humanities in Interdisciplinary Research. *Palgrave Communications, 2*, 16036.

Rau, H., Goggins, G., & Fahy, F. (2018). From Invisibility to Impact: Recognising the Scientific and Societal Relevance of Interdisciplinary Sustainability Research. *Research Policy, 47*(1), 266–276.

Rochlin, G. I. (2014). Energy Research and the Contributions of the Social Sciences: A Retrospective Examination. *Energy Research & Social Science, 3*, 178–185.

Shove, E. (2014). Putting Practice into Policy: Reconfiguring Questions of Consumption and Climate Change. *Contemporary Social Science, 9*(4), 415–429.

Sovacool, B. K., & Hess, D. J. (2017). Ordering Theories: Typologies and Conceptual Frameworks for Sociotechnical Change. *Social Studies of Science, 47*(5), 703–750.

Sovacool, B. K., Ryan, S. E., Stern, P. C., Janda, K., Rochlin, G., Spreng, D., Pasqualetti, M. J., Wilhite, H., & Lutzenhiser, L. (2015). Integrating Social Science in Energy Research. *Energy Research & Social Science, 6*, 95–99.

Stern, P. C. (2017). How Can Social Science Research Become More Influential in Energy Transitions? *Energy Research & Social Science, 26*, 91–95.

Open Access This chapter is licensed under the terms of the Creative Commons Attribution 4.0 International License (http://creativecommons.org/licenses/by/4.0/), which permits use, sharing, adaptation, distribution and reproduction in any medium or format, as long as you give appropriate credit to the original author(s) and the source, provide a link to the Creative Commons license and indicate if changes were made.

The images or other third party material in this chapter are included in the chapter's Creative Commons license, unless indicated otherwise in a credit line to the material. If material is not included in the chapter's Creative Commons license and your intended use is not permitted by statutory regulation or exceeds the permitted use, you will need to obtain permission directly from the copyright holder.

CHAPTER 10

Challenges Ahead: Understanding, Assessing, Anticipating and Governing Foreseeable Societal Tensions to Support Accelerated Low-Carbon Transitions in Europe

Bruno Turnheim, Joeri Wesseling, Bernhard Truffer, Harald Rohracher, Luis Carvalho, and Claudia Binder

Abstract Addressing global climate change calls for rapid, large-scale deployment of renewable energy technologies (RETs). Such an accelerated diffusion constitutes a new phenomenon, which challenges existing analytical approaches. The implied fundamental reconfiguration of energy systems will inevitably involve adjoining shifts in the structure of energy markets, the socio-cultural significance of energy and related rules and institutions—producing new societal tensions that are largely understudied. This chapter draws on insights from socio-technical,

---

B. Turnheim (✉)
Department of Geography, King's College London, London, UK

Manchester Institute of Innovation Research, University of Manchester, Manchester, UK

Laboratoire Interdisciplinaire Sciences Innovations Sociétés (LISIS), Université Paris-Est Marne-la-Vallée, Champs-sur-Marne, France
e-mail: bruno.turnheim@kcl.ac.uk

social-ecological and techno-economic systems studies to better understand, assess and support the exploration of low-carbon futures. We sketch out an agenda that encompasses four major tasks for governing the energy transition: i) a richer understanding of the dynamics of sociotechnical and social-ecological systems; ii) multidimensional assessments of prospective environmental, social and economic impacts of these transformations; iii) methods that enable actors to anticipate future impacts in their everyday innovation and decision practices; and iv) elaborate new governance arrangements to tackle the upcoming transformations.

Keywords Sustainability transition • Innovation • Systems • Governance challenges • Renewable energy • Interdisciplinary

J. Wesseling
Copernicus Institute of Sustainable Development, Utrecht University, Utrecht, Netherlands
e-mail: J.H.Wesseling@uu.nl

B. Truffer
Copernicus Institute of Sustainable Development, Utrecht University, Utrecht, Netherlands

Eawag—Swiss Federal Institute of Aquatic Science and Technology, Dübendorf, Switzerland
e-mail: B.Truffer@uu.nl

H. Rohracher
Department of Thematic Studies—Technology and Social Change, Linköping University, Linköping, Sweden
e-mail: harald.rohracher@liu.se

L. Carvalho
Centre of Studies in Geography and Spatial Planning, University of Porto, Porto, Portugal
e-mail: lcarvalho@letras.up.pt

C. Binder
Laboratory for Human-Environment Relations in Urban Systems, École polytechnique fédérale de Lausanne, Lausanne, Switzerland
e-mail: claudia.binder@epfl.ch

10.1 Introduction

Addressing the problems of climate change and dwindling non-renewable energy resources whilst ensuring energy security calls for the rapid and large-scale deployment of renewable energy technologies (RETs) (IEA 2015), to make up between 45% and 97% of gross final energy consumption by 2050, depending on scenarios (European Commission 2011). In order to meet the European targets, RET deployment needs to rapidly shift from early niche activities to a phase of accelerated diffusion. Since 2005, considerable progress has been made: the share of renewables is on its way to 20% and above 30% in a number of frontrunner countries (Fig. 10.1)—although there is substantial variation between countries. For technologies like solar photovoltaics (PV) or biogas, actual diffusion even significantly exceeded expectations in some countries (EEA 2017a). The higher diffusion rates have been possible thanks to a combination of ambitious targets, economic incentives (e.g. feed-in tariffs), substantial experimentation, regulatory adaptation (e.g. wind zoning laws), the emergence of industrial opportunities and the involvement of a wide range of actors.

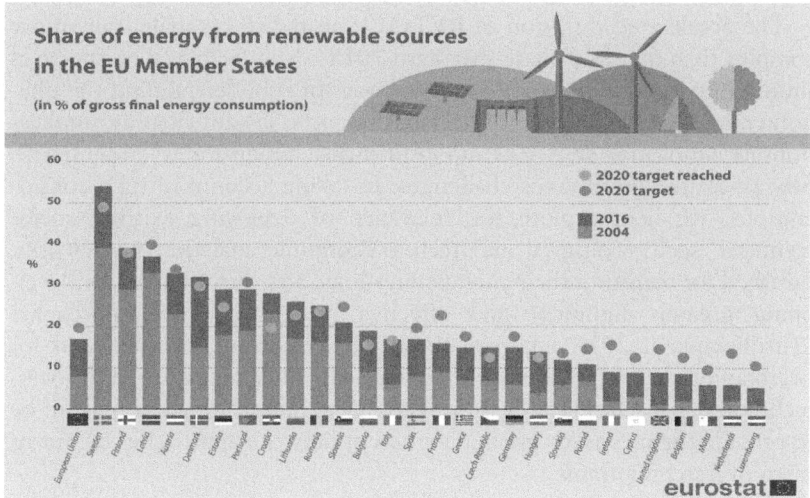

Fig. 10.1 Share of energy from renewable sources in the EU Member States. Source: Eurostat (2018)

Our core hypothesis is that as we enter this new phase of accelerated diffusion, we are presented with a new kind of phenomenon, which is characterised by different combinations of drivers and causal mechanisms (Markard 2018). Accelerated diffusion of RETs involves the transformation of existing systems, that is, mainstreaming and embedding of such technologies in society, the risk of massively disturbing existing social and natural environments, the challenging of established firms (incumbents), institutions and infrastructures. These system transformations are likely to involve tipping points (Westley et al. 2011; Olsson et al. 2006), requiring inter alia the consideration of rapidly shifting system configurations, an ability to reconsider units of analysis (e.g. from isolated technologies to systems) and core causal (innovation) mechanisms (Haydu 2010; Suurs and Hekkert 2009). For instance, it may require shifting our focus from the development and deployment for specific technologies (e.g. solar PV, off-shore wind) to questions of interactions, system integration and reconfiguration of whole electricity systems, implying different challenges for governance. The extant analytical frameworks that deal with the emergence of new technologies seem not well prepared for this task, as most research to date has focused on the early phases of RET diffusion, which do not generate deep impacts on overall energy system configurations and are relatively inoffensive to established actors.

The accelerated diffusion of RETs is expected to be analytically more complex than the early phase (Markard 2018). New analytical perspectives should in particular inform about new ways of i) understanding the phenomenon at hand, ii) assessing related impacts, iii) anticipating implications for innovation strategies and institutional design and iv) dealing with new associated governance challenges. In taking account of such considerations, we here explore the relevance of deploying existing sociotechnical, socio-ecological and techno-economic analytical frameworks, the need for revisiting their core assumptions, and the potential for developing greater alignment and effective bridges between approaches (Turnheim et al. 2015; Cherp et al. 2018). We posit that a crucial task for overcoming inevitable blind spots of any individual approach (e.g. sociotechnical approaches lack detail of ecological impact dimensions) will be to provide the means for greater alignment between approaches by means of an overarching frame.

Concerning the development of such an overarching interdisciplinary frame, we consider needs for adjustments (within specific approaches) and disciplinary integration (across approaches). We specifically attend to the following questions:

- *Are existing frameworks fit for purpose in this new phase of accelerated diffusion?*
- *Do they address the core mechanisms of the new phase?*
- *And if not, how can they be adjusted or complemented with different perspectives?*

In this chapter, we provide tentative answers to such explorative questions and provide implications for policy and practice in dealing with accelerated RET diffusion.

10.2 What Does RET Accelerated Diffusion Look Like and How Can We Make Sense of It?

Systems perspectives are crucial for understanding the successful development, implementation and accelerated diffusion of new technologies, because the success of this transformative process depends on a wide range of interacting social, economic, technological and environmental factors (EEA 2017b). We consider three relevant but distinct analytical approaches as starting points:

- Techno-economic systems approaches provide the most conventional frame for the study of system change (e.g. in quantitative models and scenarios) but tend to favour technological substitution patterns and neglect reconfigurational change and its unfolding over time.
- Socio-technical approaches emphasise system interactions relevant to innovation dynamics and their governance, rooted in co-evolutionary understandings of change, but tend to be less detailed on specific environmental impacts.
- Socio-ecological approaches problematise interactions of social structures and environmental systems, conceptualising change in terms of dynamic equilibria and tipping points, but tend to be less informative about how systemic change can be achieved.

In the acceleration phase, the new socio-technical systems of RETs undergo rapid change and lead to a multitude of impacts both on environmental and socio-economic dimensions. As a consequence, we need to better understand the new quality of the underlying processes. Table 10.1 maps out the kinds of processual shifts that can be observed between i) an

Table 10.1 Differences in formative and reconfiguration phase of energy systems change across multiple dimensions

	Formative/niche phase	*Acceleration/regime reconfiguration—signs of stress*
Incumbents positioning towards the new technologies	Incumbent actors' discourse is characterised by a combination of a) ignoring, downplaying or discrediting emerging transformative discourse and b) acknowledging emerging issues and seeking to shape broad discourses Incumbent actors marginally explore alternative options (e.g. RETs) as part of early portfolio diversification strategies Industry associations show a closed industry front	Incumbent actors may show signs of erratic behaviour, strategic reorientation and divergence (heterogeneity) Incumbent actors' strategies become increasingly characterised by ambidexterity: focusing on exploiting established options and exploring new portfolios more systematically (e.g. buying up successful new entrants, integrating in-house projects) Industry associations are increasingly stretched due to conflicting and diverging interests of their members Accordingly, more specialised challenger associations become prominent
New entrants positioning towards the new technologies	New entrants introduce and trial out a wide range of new options and technologies (no clear dominant design) New discourse coalitions and advocacy alliances emerge, generating significantly coherent alternative narratives (e.g. broad RET agenda) Broad agendas emphasising societal benefits (e.g. climate mitigation) allow the coexistence of a variety of options and visions (including radical visions)	New entrants experience difficulties with scaling up business (e.g. in terms of finance, customer base, logistics), signalling a need for new business models New entrants shift emphasis on economic attractiveness over societal benefits New entrants increasingly engage with mainstreaming and standardisation (which can lead to a watering down of initial ambitions and radical visions) Alternative visions become increasingly specific (e.g. related to particular options such as solar PV), resulting in the possible emergence of conflicts

(*continued*)

Table 10.1 (continued)

	Formative/niche phase	Acceleration/regime reconfiguration—signs of stress
Policy	Development and deployment policies focus on generating support for R&D and learning and developing new market niches	Existing policies lag behind the realities of accelerated diffusion, resulting in a mismatch between policy instruments and their goals (e.g. German feed-in tariffs are being portrayed as cross-subsidies for Chinese PV production, urgency of sunset clauses) Increased policy expenses may erode legitimacy; controversies arise over policy adaptation and reorientation Unintended effects of early support policies generate increasing legitimacy challenges prompting review, refocus or policy phase-out New policy issues emerge: standardisation support, system integration, dealing with incumbents, dealing with potential losers and conflicts
Markets and legitimation	(Proto-) markets are actively constructed by RET proponents Wide price gaps between established and renewable energy technologies Public raise questions about the value and relevance of RETs	Users become increasingly familiar with RETs Declining costs, cost differential between conventional energy and RETs narrow down RETs become gradually legitimated and mainstreamed
Technology	Focus on radical innovation. Diverse offer of technical solutions to societal problems	Focus on incremental improvements Complementary innovations and increasing interactions between RETs Selection of innovations and embedding in society Focus on shaping system architectures (e.g. new structure of transmission grids) and supporting arrangements

(*continued*)

Table 10.1 (continued)

	Formative/niche phase	Acceleration/regime reconfiguration—signs of stress
Infrastructure	Experiments at different scales with radically new infrastructure solutions (e.g. smart grids); large-scale incremental change upgrades to infrastructure (e.g. more, bigger and longer transmission lines)	Unanticipated problems emerge, such as unstable grids (see effect of RETs on Spanish grid stability) Investments in novel infrastructures go to scale (e.g. rapid roll-out of charging points for electric vehicles)
Research	Uncertainty and disagreement on what the new dominant technology will be (e.g. electric vehicles vs. hydrogen fuel cell vehicles over past 15 years)	Convergence on (cognitive) expectations and visions for the future (as solutions get selected)
Finance and other resources	Exploring new business model innovations (e.g. renting out energy-saving appliances or RETs). Many start-ups are able to get funding for initial experimentation and market introduction	Many start-ups have problems getting finance for independent scaling up and are acquired by incumbents. Human resources challenges: need for reskilling workforce (from fossil fuel to RET) Need to mobilise significant capital for scaling up
Ecological dynamics and environmental impacts	Benign interactions and marginal stresses Low consideration of trade-offs	Increasing pressure on environmental dimensions for RETs: acute environmental impacts and potential tipping points; dwindling ecological systems and potential loss of ecosystem functions; and acute competition for space and land Pressing need to consider trade-offs (e.g. biomass and food, PVs and mineral extraction) Experimentation with new arrangements to handle impacts (e.g. zoning for wind, limitations for certain kinds of biomass)

early formative phase of diffusion, characterised by experimentation and the formation of niche markets that may require R&D support and measures to protect alternatives from mainstream selection environments, and ii) an acceleration phase, characterised by the rapid scaling of RET diffusion and their integration into larger technical, societal and environmental systems—in paths that are yet to stabilise. Table 10.1 focuses specifically on identifying the signs of stress that we are likely to evidence in such processual shifts (from early stages of socio-technical diffusion to accelerated diffusion and from benign socio-ecological interactions to increasing stresses and pressures). Mapping these signs of stress against current developments indicates that in many cases we have entered this new acceleration phase.

The impact of the RET transition on different environmental and socio-economic dimensions and the associated governance challenges will become much clearer as the acceleration phase progresses. However, the implications of the Collingridge dilemma (Collingridge 1980) become apparent in this case and lend some urgency to better understand the dynamics as they unfold: at first the transition is still malleable and can be steered, but information about in which direction to steer in is limited as the potential consequences of the different transition pathways remain unclear. As some transition pathways are abandoned, and others gain momentum and become embedded in society, the consequences become clear, but the pathways are more difficult to shape due to multiple sources of lock-in (Klitkou et al. 2015).

Hence, while this acceleration phase is a typical 'hot phase' characterised by disruption, high uncertainty and fluidity (Callon 1998), it is a decisive moment in which the overall direction of change is likely to be settled, with implications on how the transition will unfold and what kind of system we will end up with. Consequently, a reflexive attitude towards the impacts of disruption and emergent governance challenges is key, so that we can anticipate and adequately guide the transition process at this critical determining point in time, after which we are likely to witness a new phase of stabilisation and lock-in. Influencing these new forms of lock-in becomes a relevant undertaking.

10.3 Do Existing Frameworks and Policies Suffice?

Existing analytical and associated policy approaches like socio-technical and socio-ecological frameworks say little about the specific mechanisms at play in the acceleration phase. Despite a number of historical case studies covering entire (energy) transitions (see Martínez Arranz (2017) for a review), transitions studies have primarily focused on how the formative phases of energy transitions can be stimulated. Phases of rapid regime reconfigurations have gained much less attention also because they have only recently started to appear in empirical reality. The question that emerges is: 'How can new frameworks be developed that are able to account for the inherent uncertainty, turbulence, conflicts, struggles playing out in this disruptive phase?'

Within the socio-technical literature, different systems perspectives can be identified. The Multi-Level Perspective is useful for describing the overall characterisation of transitions dynamics as the interplay between exogenous pressures and forces of creative destruction emerging in protected spaces that put pressure on the established technologies and infrastructures that provide societal functions like energy provision. But it tends to overlook the micro-level mechanisms underpinning specific diffusion processes. These processes may be better captured by the Strategic Niche Management framework that focuses on the role of visions, learning and building social networks in the development and accumulation of niches (Schot and Geels 2008). The accelerated diffusion phase has mostly been conceptualised as a problem of stacking policy-protected niche markets (called niche accumulation). Recent developments about *niche empowerment* consider issues of wider embedding (Raven et al. 2015). The technological innovation systems approach provides an explicit stage model of the maturation of novel technologies and products. It emphasises core processes that come to bear in technology maturation and market expansion (Bergek et al. 2008; Suurs and Hekkert 2009). Finally, the more generic concept of transformative system failures by Weber and Rohracher (2012) are a fruitful starting point for understanding the dynamic (governance) challenges of transitions but do not differentiate between the stages of system development. Particularly useful for dealing with the uncertainty of acceleration is the concept of branching points as it can suggest where/when opportunities for directional governance may be expected and focus attention where reflexivity regarding impacts is most critical (Rosenbloom et al. 2018).

Research on socio-ecological systems come in two guises: the Natural Science approach uses concepts like tipping points to assess the global impact of RET transition on the planetary boundaries of the Earth system. The Social Science approach is more solution-driven, locally oriented and, like the socio-technical systems field, incorporates actors, institutions, networks and infrastructure. Reviewing the socio-ecological systems literature, O'Brien et al. (2017) identify three main analytical approaches:

- resilience approaches that build on ecological understanding
- pathways approaches that outline different trajectories to meeting sustainability goals
- spheres of transformation approaches that highlight the practical, political and personal domains for effectuating transformation

Although socio-ecological approaches are useful for understanding the social and environmental impacts of accelerated RET diffusion, their major drawback lies in their inability to explain how desirable systemic change could be achieved (*Ibid*). SES could profit from considering insights from the STS literature dealing with transitions. The other way around, recent studies have suggested that transitions research could also be inspired by SES concepts, such as the resilience of transition pathways, for better characterising and steering the acceleration phase. The idea is that the transition process itself, having a normative goal in the energy transition, should be resilient, that is, should be able to continue on the pathway even if major changes in the overall policy environment occur (Binder et al. 2017).

Within techno-economic approaches, Integrated Assessment Models have proven useful by providing quantitative models that incorporate technical, economic and social factors to predict long-term (2050 and 2100) impacts on climate change, resources and biodiversity (van Vuuren and Hof 2017). Notably, it was the Integrated Assessment Models that most clearly indicated the certainty of catastrophic climate change impacts in the absence of drastic policy interventions (Cherp et al. 2018). One of the drawbacks of Integrated Assessment Models is however that their mathematical functions assume a relatively smooth RET diffusion and do not take into account 'major or abrupt shocks, tipping points or any other non-linear system behaviour' (EEA 2017b, pp. 14–15) which typically characterise adoption and transition processes (Geels and Schot 2007; Rogers 2003).

To conclude, although the socio-technical, socio-ecological and techno-economic perspectives continue to develop and borrow from each other, they warrant further conceptual development to better understand, assess and deal with the new governance challenges associated with the accelerated diffusion of RETs.

10.4 Implications for Policy and Practice

Reflecting on what the different systems literatures have uncovered on the transition to RETs so far, we provide some preliminary warnings for policy, practice and research.

- *Act now:* First, the transition to RETs is happening at an increasingly rapid pace at global, European, country and regional scales. As momentum increases, the window of opportunity for steering the transition process and its various local components in a given direction is closing. Since it will be increasingly difficult to shape the further development the farther the scaling has progressed, policy should provide clear, long-term goals while remaining flexible enough to acknowledge and accommodate the inherent uncertainties of societal transitions.
- *Target the whole system and all stages of the innovation process:* Providing such normative directionality means that existing policies and governance structures need to be adapted to adequately steer the transition process. This means moving beyond conventional innovation policies directed towards R&D and including multi-stakeholder governance arrangements and demand-side instruments that reward the uptake of renewable energy-related technologies and penalise polluting options.
- *Involve different stakeholders in reflexive governance:* To support good decisions, it is important to critically assess the different social and environmental impacts on the system and further open up normative discussions by involving different stakeholder groups. These stakeholders should be approached equally, lest the discussion is captured by the power of vested interests. Reflexivity regarding the direction of the transition, emerging impacts and societal goals remains crucial during the phase of acceleration, particularly when critical choices need to be made at transition branching points (cf. Rosenbloom et al. 2018). The concept of resilience of transition might provide a good starting point for policy development.
- *Consider how to overcome path dependencies and vested interests:* Furthermore, it is important to note that as the transition progresses and RET diffusion accelerates, decisions regarding directionality become increasingly political as their impact increases. Vested interests become more seriously threatened by new economic powers,

and incumbents have to increasingly commit to and select within the new options, while abandoning the old. They have a history of opposing and shaping political decisions that are not in their favour and although opposition becomes increasingly illegitimate, shaping endeavours may increase as the stakes increase (Wesseling et al. 2014). Closed industry fronts of opposition to change can be broken by engaging with individual, innovative companies instead of industry associations that typically prefer the status quo (*Ibid.*).

- *Deal with the potential losers of transition:* The transition will ultimately happen and there will be losers. It is important to acknowledge this and the fact that doing nothing means favouring the status quo and can have even higher societal costs on the long-term than timely adaptation. Instead, policy should proactively deploy strategies to deal with losers, for example, develop re-education schemes for those currently employed in those sectors that are to be phased out.
- *Account for differences in transitions across time and space:* Given that diffusion and system integration dynamics unfold at different speeds and in qualitatively different ways across countries and regions (e.g. around different interests and trade-offs, due to different resource endowments, strength of local coalitions and ante-coalitions), the issue of dealing with such variations becomes a new priority that is especially salient at for European governance.

10.5 How to Proceed?

To systematically approach the governance challenges associated with accelerated RET diffusion, we distinguish four analytical challenges that the aforementioned systems literatures will have to tackle in the future:

1. *Understanding system dynamics*: First it is important to develop a better understanding of the potential socio-technical/socio-ecological dynamics of the new acceleration phase, as literature has done for the formative stage of the sustainable energy transition (see Table 10.1). This requires an inventory of the recent contributions in the different systems literatures that shed insights in the explosive dynamics of this stage, such as concepts like branching points, system resilience, tipping points and so on, and explore cross-fertilisation across these literatures to develop new or existing frameworks.

2. *Assessing signs of systemic stress:* This deeper understanding should then inform assessment of the social, environmental (ecological and space) and economic impacts of accelerated RET diffusion. One way of doing so is by providing signs of 'stress' that indicate acceleration and critical decision making at branching points.
3. *Anticipating future social and ecological impacts:* New understandings about system dynamics and their impacts should inform individual actors in anticipating the future impacts of their decisions made today. Examples of anticipatory approaches include value-based designs and constructive technology assessment (Truffer et al. 2017).
4. *Transforming systems and their governance:* Other than informing individual actors, new understandings about system dynamics and their impacts should also inform system-level governance structures to steer the direction and rate of the transition process. An example of such an approach is transition management (Kemp et al. 2007), although it has so far focused on the formative stage of transition. These approaches should direct system transformation on the basis of the social, environmental and economic impacts of different transition pathways, which is currently overlooked in particularly the socio-technical literature (Kemp and Van Lente 2011).

10.6 Final Reflections

The increased rate of deployment of RETs is a welcome sign of progress towards low-carbon transitions. It comes with new challenges that this chapter has sought to highlight. Our core hypothesis suggests that as we enter this new phase, we are confronted with a qualitatively different phenomenon that warrants a new reflection concerning the appropriateness of current analytical and governance approaches. The complexities, uncertainties, temporal and political issues involved need to be more centrally recognised as the keys to effective and legitimate interventions. The increased engagement of a variety of Social Science perspectives with core-related issues is a significant strength to draw on, continuously improve and cross-fertilise (Castree et al. 2014; Cherp et al. 2018; Kuzemko et al. 2016; Stirling 2014; Turnheim et al. 2015). A new wave of interdisciplinary research is emerging that explicitly recognises the task at hand. This chapter has built on and contributed to this collective effort by charting a possible way forward.

REFERENCES

Bergek, A., Jacobsson, S., Carlsson, B., Lindmark, S., & Rickne, A. (2008). Analyzing the Functional Dynamics of Technological Innovation Systems: A Scheme of Analysis. *Research Policy, 37*(3), 407–429.

Binder, C. R., Mühlemeier, S., & Wyss, R. (2017). An Indicator-based Approach for Analyzing the Resilience of Transitions for Energy Regions. Part I: Theoretical and Conceptual Considerations. *Energies, 10*(1), 36.

Callon, M. (1998). An Essay on Framing and Overflowing: Economic Externalites Revisted by Sociology. In M. Callon (Ed.), *The Laws of the Market* (pp. 244–269). Oxford: Blackwell Publishers/The Sociological Review.

Castree, N., Adams, W. M., Barry, J., Brockington, D., Buscher, B., Corbera, E., Demeritt, D., Duffy, R., Felt, U., Neves, K., Newell, P., Pellizzoni, L., Rigby, K., Robbins, P., Robin, L., Rose, D. B., Ross, A., Schlosberg, D., Sorlin, S., West, P., Whitehead, M., & Wynne, B. (2014). Changing the Intellectual Climate. *Nature Climate Change, 4*, 763–768.

Cherp, A., Vinichenko, V., Jewell, J., Brutschin, E., & Sovacool, B. (2018). Integrating Techno-economic, Socio-technical and Political Perspectives on National Energy Transitions: A Meta-theoretical Framework. *Energy Research & Social Science, 37*, 175–190.

Collingridge, D. (1980). *The Social Control of Technology*. New York: St. Martin.

EEA. (2017a). *Renewable Energy in Europe 2017 Recent Growth and Knock-on Effects—European Environment Agency, Report N0 3/2017*. Copenhagen: European Environment Agency.

EEA. (2017b). *Perspectives on Transitions to Sustainability—European Environment Agency. Report No 25/2017*. Copenhagen: European Environment Agency.

European Commission. (2011). *Energy Roadmap 2050*. Brussels: European Commission.

Eurostat. (2018). *Renewable Energy Statistics*. Based on data Extracted in June 2018. [online]. Retrieved June 28, 2018, from http://ec.europa.eu/eurostat/statistics-explained/index.php/Renewable_energy_statistics

Geels, F. W., & Schot, J. (2007). Typology of Sociotechnical Transition Pathways. *Research Policy, 36*(3), 399–417.

Haydu, J. (2010). Reversals of Fortune: Path Dependency, Problem Solving, and Temporal Cases. *Theory and Society, 39*(1), 25–48.

IEA. (2015). *Energy Technology Perspectives 2015: Mobilising Innovation to Accelerate Climate Action*. Paris: International Energy Agency.

Kemp, R., & Van Lente, H. (2011). The Dual Challenge of Sustainability Transitions. *Environmental Innovation and Societal Transitions, 1*(1), 121–124.

Kemp, R., Loorbach, D., & Rotmans, J. (2007). Transition Management as a Model for Managing Processes of Co-evolution Towards Sustainable Development. *International Journal of Sustainable Development & World Ecology, 14*(1), 78–91.

Klitkou, A., Bolwig, S., Hansen, T., & Wessberg, N. (2015). The Role of Lock-in Mechanisms in Transition Processes: The Case of Energy for Road Transport. *Environmental Innovation and Societal Transitions, 16*, 22–37.

Kuzemko, C., Lockwood, M., Mitchell, C., & Hoggett, R. (2016). Governing for Sustainable Energy System Change: Politics, Contexts and Contingency. *Energy Research & Social Science, 12*, 96–105.

Markard, J. (2018). Implications for Research and Policy. *Nature Energy.* doi:https://doi.org/10.1038/s41560-018-0171-7.

Martínez Arranz, A. (2017). Lessons from the Past for Sustainability Transitions? A Meta-analysis of Socio-technical Studies. *Global Environmental Change, 44,* 125–143.

O'Brien, K., Sygna, L., Datchoua, A., Pettersen, S., & Rada, R. (2017). 2. Transformations in Socio-ecological Systems. In M. Asquith, J. Backhaus, F. Geels, A. Golland, A. Hof, R. Kemp, T. Lung, K. O'Brien, F. Steward, T. Strasser, L. Sygna, D. van Vuuren, & P. Weaver (Eds.), *Perspectives on Transitions to sustainability—European Environment Agency, Report No. 25/2017.* Copenhagen: EEA.

Olsson, P., Gunderson, L. H., Carpenter, S. R., Ryan, P., Lebel, L., Folke, C., & Holling, C. S. (2006). Shooting the Rapids: Navigating Transition to Adaptive Governance of Social-ecological Systems. *Ecology and society, (1),* 11, 18.

Raven, R., Kern, F., Verhees, B., & Smith, A. (2015). Niche Construction and Empowerment Through Socio-political Work. A Meta-analysis of Six Low-carbon Technology Cases. *Environmental Innovation and Societal Transitions, 18,* 164–180.

Rogers, E. (2003). *Diffusion of Innovations.* New York: Simon and Schuster.

Rosenbloom, D., Haley, B., & Meadowcroft, J. (2018). Critical Choices and the Politics of Decarbonization Pathways: Exploring Branching Points Surrounding Low-carbon Transitions in Canadian Electricity Systems. *Energy Research & Social Science, 37*, 22–36.

Schot, J., & Geels, F. W. (2008). Strategic Niche Management and Sustainable Innovation Journeys: Theory, Findings, Research Agenda, and Policy. *Technology Analysis & Strategic Management, 20*(5), 537–554.

Stirling, A. (2014). Transforming Power: Social Science and the Politics of Energy Choices. *Energy Research & Social Science, 1,* 83–95.

Suurs, R. A. A., & Hekkert, M. P. (2009). Cumulative Causation in the Formation of a Technological Innovation System: The Case of Biofuels in the Netherlands. *Technological Forecasting and Social Change, 76*(8), 1003–1020.

Truffer, B., Schippl, J., & Fleischer, T. (2017). Decentering Technology in Technology Assessment: Prospects for Socio-technical Transitions in Electric Mobility in Germany. *Technological Forecasting and Social Change, 122,* 34–48.

Turnheim, B., Berkhout, F., Geels, F., Hof, A., McMeekin, A., Nykvist, B., & van Vuuren, D. (2015). Evaluating Sustainability Transitions Pathways: Bridging

Analytical Approaches to Address Governance Challenges. *Global Environmental Change, 35*, 239–253.
van Vuuren, D., & Hof, A. (2017). 6. Integrated Assessment Modelling Approaches to Analysing Systemic Change. In M. Asquith, J. Backhaus, F. Geels, A. Golland, A. Hof, R. Kemp, T. Lung, K. O'Brien, F. Steward, T. Strasser, L. Sygna, D. van Vuuren, & P. Weaver (Eds.), *Perspectives on Transitions to Sustainability—European Environment Agency, Report No. 25/2017*. Copenhagen: EEA.
Weber, K. M., & Rohracher, H. (2012). Legitimizing Research, Technology and Innovation Policies for Transformative Change: Combining Insights from Innovation Systems and Multi-Level Perspective in a Comprehensive 'Failures' Framework. *Research Policy, 41*(6), 1037–1047.
Wesseling, J. H., Farla, J. C. M., Sperling, D., & Hekkert, M. P. (2014). Car Manufacturers' Changing Political Strategies on the ZEV Mandate. *Transportation Research Part D: Transport and Environment, 33*, 196–209.
Westley, F., Olsson, P., Folke, C., Homer-Dixon, T., Vredenburg, H., Loorbach, D., Thompson, J., Nilsson, M., Lambin, E., Sendzimir, J., Banerjee, B., Galaz, V., & van der Leeuw, S. (2011). Tipping Toward Sustainability: Emerging Pathways of Transformation. *Ambio, 40*(7), 762–780.

Open Access This chapter is licensed under the terms of the Creative Commons Attribution 4.0 International License (http://creativecommons.org/licenses/by/4.0/), which permits use, sharing, adaptation, distribution and reproduction in any medium or format, as long as you give appropriate credit to the original author(s) and the source, provide a link to the Creative Commons license and indicate if changes were made.

The images or other third party material in this chapter are included in the chapter's Creative Commons license, unless indicated otherwise in a credit line to the material. If material is not included in the chapter's Creative Commons license and your intended use is not permitted by statutory regulation or exceeds the permitted use, you will need to obtain permission directly from the copyright holder.

CHAPTER 11

Towards a Political Ecology of EU Energy Policy

Gavin Bridge, Stefania Barca, Begüm Özkaynak, Ethemcan Turhan, and Ryan Wyeth

Abstract At the root of energy policy are fundamental questions about the sort of social and environmental futures in which people want to live and how decisions over different energy pathways and energy futures are made. The interdisciplinary field of political ecology has the capacity to address such questions, while also challenging how energy

G. Bridge (✉) • R. Wyeth
Department of Geography, Durham University, Durham, UK
e-mail: g.j.bridge@durham.ac.uk; ryan.d.wyeth@durham.ac.uk

S. Barca
Center for Social Studies, University of Coimbra, Coimbra, Portugal
e-mail: sbarca@ces.uc.pt

B. Özkaynak
Department of Economics, Boğaziçi University, Istanbul, Turkey
e-mail: begum.ozkaynak@boun.edu.tr

E. Turhan
Environmental Humanities Lab, Division of History of Science, Technology and Environment, KTH Royal Institute of Technology, Stockholm, Sweden
e-mail: ethemcan@kth.se

© The Author(s) 2018
C. Foulds, R. Robison (eds.), *Advancing Energy Policy*,
https://doi.org/10.1007/978-3-319-99097-2_11

policy conventionally gets done. We outline a political ecology perspective on EU energy policy that illuminates how the distribution of social power affects access to energy services, participation in energy decision-making and the allocation of energy's environmental and social costs.

Keywords Energy transitions • Political ecology • Knowledge • Scale • Democracy • Eco-sufficiency • Justice

11.1 Introduction

This chapter outlines a political ecology perspective on EU energy policy. Political ecology is an interdisciplinary approach to understanding and transforming human-environment relations. It focuses on how economic and political power shape social and environmental outcomes and is informed by both critical social theory and the experience of social movements. Political ecology is a reflexive (i.e. 'self-conscious') form of knowledge production. It pays close attention to how hegemonic power is sustained through scientific concepts and popular discourses around management of society-environment relations (e.g. scarcity, security, efficiency and risk). It also unsettles and problematises dominant forms of knowledge by generating alternative data and concepts, often through research on and with marginalised social groups.

We show in this chapter how a political ecology perspective not only asks different questions about energy policy but also poses a challenge to how energy policy traditionally has been done. Our account draws together several insights from political ecology research which, to date, has focused more on environmental policy and governance than energy per se. The political ecology perspective we offer involves grounded, empirically based assessment of how social power affects access to energy services, participation in energy decision-making and allocation of energy's environmental and social costs. It also encompasses a broader 'ecology of politics' (Huber 2015) that examines how the histories and geographies of energy stocks and flows reproduce social power (i.e. dominance and vulnerability) at a range of spatial scales.

11.2 A Political Ecology Perspective

Political ecology has deep and multiple roots. It draws in equal measure on critical social theory, historical materialism and the experience and knowledge of social movements seeking to redress historical patterns of social and environmental injustice. It coalesced as a recognisable body of thought in the 1970s and 1980s, as a critical response to technocratic and managerialist approaches to the environment and the obsession at the time with issues like overpopulation, resource scarcity and the carrying capacity of the Earth (Bridge et al. 2015). Its provocative coupling of two words from different traditions of thought directly challenges the supposedly 'apolitical' character of expert environmental management (Robbins 2011; M'Gonigle 1999). Political ecologists argue that mainstream scientific and managerial approaches to the environment fail to adequately question existing socio-economic arrangements, such as relations around gender, class and race, and historic patterns of dominance and marginalisation at different geographical scales. Consequently, they overlook the root causes of apparently 'environmental' problems which, political ecologists argue, are to be found in the unequal distribution of power within society. In this way, political ecology casts critical light on how conventional scientific and management approaches, through claims about expertise and scientific objectivity, often work to advance the interests of dominant classes and social groupings while keeping others marginalised. Political ecology offers, therefore, both an alternative account of the origins of environmental problems and a critique of the knowledge frameworks through which those problems are apprehended and solutions defined. Political ecology is a form of *praxis*—a unity of theory and practice orientated towards social change—and gives researchers a toolbox for critical and engaged analysis (Loftus 2017).

To date, there is little research on (EU) energy policy from a political ecology perspective. Researchers can draw, however, on two primary insights from political ecology's substantial record of work on environmental conflicts. First, political ecology highlights how flows of energy and raw materials ('socio-metabolism') create the conditions of possibility for economic and political power at a range of scales, from the geopolitics of international trade to relations of responsibility, autonomy and identity associated with energy consumption and citizenship (Huber 2015). It illuminates how social values, knowledge and political organisation have co-evolved with growing energy consumption and how energy transition involves not only substituting fuels or improving energy efficiency but also considering how energy systems and infrastructures create different political possibilities. Second, political ecology shows how the socio-political context of knowledge production shapes perceptions of the problem at hand and how this 'situated' character of knowledge influences the choices available for addressing and managing matters of concern. Political ecology breaks down the 'knowledge silos' of traditional economic or technical analyses (a feature it shares with other interdisciplinary initiatives, like sustainability science) but also challenges powerful hierarchies around assumed expertise: it highlights how calls for interdisciplinarity often overlook the wealth of 'lay' knowledge among those who live and work in and around sites of environmental crisis and conflict. In this way, political ecology expands the range of voices heard when researching energy and environmental policy issues, offering a distinctive 'view from below'. The alternative geographies, scales and histories originating from the experience of affected communities and environmental justice organisations can significantly enrich—and transform—policy analyses (Temper et al. 2018). Empirical findings and conceptual perspectives originating in these communities—with prominent energy-related examples include ecological debt, climate justice and degrowth—can be mobilised at regional, national and international levels to press for more ethical forms of public decision-making (Martinez-Alier et al. 2014).[1]

11.3 An Alternative Lens on EU Energy Policy

Conventional accounts of EU energy policy tell the story of policy trajectories 'from above'. They are contemporary versions of 'Chevalier's Dream', the century-long aspiration of building a modern Europe by 'Eradicating poverty, achieving independence from nature, and creating lasting peace' (Högselius et al. 2015). Most accounts focus, for example, on delivering an EU Energy Strategy and Energy Union that ensures 'secure, competitive and sustainable energy', integrating energy infrastructures through cross-border construction and harmonising network codes, expanding EU competencies in energy policy over time and/or unresolved scalar tensions between national interests and supranational objectives. What political ecology offers, in this context, is an alternative lens on the 'problems' at the heart of energy policy in the EU. This lens reveals some of the unspoken assumptions underpinning current energy policy and strategy, highlights how they limit possibilities for action and invites us to reformulate policy in different ways. Here we outline three such alternatives.

11.3.1   Towards Energy Sufficiency: Beyond Economic Growth and Ecological Modernisation

The historical materialist perspective at the heart of political ecology enables a critical reappraisal of mainstream narratives about Europe's past energy transitions, now embedded in political choices that present themselves as being in the interest of 'the people'. A core storyline about energy transition in Europe centres on the enormity of the energy leap that (western) European countries made after the Industrial Revolution (Kander et al. 2013). Once upon a time, the story goes, Europe was constrained by the scarcity of its natural resources relative to population. However, fossil fuels—coal first, then oil and natural gas—allowed Europe to escape this trap, grow rich and become a dominant force in the world economy. In this storyline fossil fuels were a necessary precondition for modern economic growth (MEG), where the term 'modern' implies simultaneous increases in population and per capita income (Barca 2011). More recently, a second storyline complements the core MEG narrative underpinning EU energy policy: ecological modernisation (EM). This centres on decreasing energy consumption per unit of GDP in the industrialised countries of western Europe, emphasising how this pattern, once generalised to developing countries, will lead to decarbonisation of the world economy (White et al. 2016). EM is now embedded in EU energy and environmental policies and in global climate policy, despite its shortcomings.[2] Together, MEG-EM storylines shape three important assumptions underpinning EU energy policies: that (1) growing levels of energy consumption are socially necessary (underpinning concerns about *security* of supply), (2) energy must be cheap to fuel economic growth (the significance of *affordability*) and (3) growing energy consumption can be compensated by 'dematerialising' the economy (the attention to *decarbonising* the energy sector). These assumptions are reflected in the EU Energy Strategy's top-level objective of ensuring 'secure, competitive and sustainable energy', as highlighted above.

A political ecology perspective on Europe's energy transition is premised on quite different narratives. Informed by studies of social and environmental history, political ecologists have studied the social, spatial, gender and environmental inequalities arising from MEG and EM processes, showing how Europe's energy transitions have been achieved through a global process of unequal exchange. For example, the first industrial revolution—centred on textiles—involved appropriating time (labour) and space (land) associated with cotton and wool production outside Europe and displacing the environmental loads of fibre production to overseas colonies (Hornborg 2006). Similarly, the partial decarbonisation of (northern) European economies today is due to deindustrialisation and the relocation of carbon-intensive production elsewhere (Bumpus and Liverman 2008). Political ecology identifies how MEG and EM have given rise to ecological distribution conflicts and to struggles around knowledge, risk and precaution in the face of scientific/technical uncertainties and for the recognition of rights and participation claims (Martinez-Alier et al. 2010). Historical research in political ecology, for example, has brought to light the key role of grassroots anti-nuclear mobilisation in southern Europe during the 1970s and 1980s, overlooked by previous research because it did not correspond to the postmaterialist model of 'new social movements' postulated by Political Science (Barca and Delicado 2016). Research has also given a critical account of the high-risk politics of hydropower in Italy and Spain, as driven by powerful economic interests with disregard for the vernacular knowledge and safety of local communities (Huber et al. 2016).

Work like this can reformulate the goals of EU energy policy. Instead of pursuing cheap, secure and clean energy, it steers attention towards eco-sufficiency and prospects for degrowth. The former implies reducing consumption to ensure equal access to sufficient means of production within the limits of ecological reproduction (Salleh 2009); the latter posits all societies, starting with the wealthiest, should disengage from practices that accelerate the throughput of energy and resources (Petridis et al. 2017). Degrowth and eco-sufficiency offer striking alternatives to the policy triplet of 'secure, competitive and sustainable energy'. They prioritise reductions in consumption in addition to pursuing 'clean energy' (a strategy that, on its own, legitimises land and water grabbing) and consider energy a social 'commons' to be shared, rather than secured and commodified. As a consequence, degrowth and eco-sufficiency challenge institutional and cultural practices around energy at both supranational (EU) and national levels.

11.3.2 From Consumers to Citizens: An Expanded Sense of Identity and Demands

Political ecology's grounded and 'bottom-up' approach to formulating the problems and solutions that lie at the heart of energy policy reveals a repertoire of identities, perceptions and demands. It exposes the mythical figure of the 'average consumer' that permeates EU energy policy and highlights how EU citizens have multiple demands for energy system change that exceed those of decarbonising, securing and making energy more competitive. Political ecology identifies the multiple reasons people protest and resist, the 'communities' of shared experience that form around energy infrastructures and the way these communities give voice to a rich set of alternative imaginaries (see Genus et al., Chap. 9 in this collection, for definition) around energy provision that include calls for responsibility, autonomy and sovereignty. A key demand from citizens centres on energy democracy—the anti-nuclear and anti-fracking movements are examples—so that, when it comes to 'power to the people', it is voice rather than kilowatts that people demand (Burke and Stephens 2017). Communities frequently draw a clear link between distributional concerns (e.g. environmental health and security) and claims for recognition (the defence of basic human rights and territorial rights) and/or participation in decision-making. For example, communities challenging energy projects—such as the Trans Adriatic Pipeline (TAP) supported by the European Investment Bank[3]—often face police violence and have their concerns dismissed as 'NIMBYism'. Political ecology takes seriously the demands of these place-based social movements and their capacity for envisioning new transition pathways that promote environmental sustainability and social justice. Communities that form around energy infrastructure and energy policy are not necessarily progressive: infiltration of the renewable energy sector by mafia groups, profiting from subsidies available exclusively to domestic users and farmers (Caneppele et al. 2013) or facilitating landgrabs, underlines the importance of focusing on power relations and structural inequalities while enabling a more people-centred and democratic energy system.

A closer look at conflicts around EU-related energy projects indicates the role such struggles might play in guiding energy choices. The map of the imagined community of 'Blockadia' in the Environmental Justice Atlas is a case in point: compiled by a network of political ecology researchers, it brings together worldwide cases of people defending their land, livelihoods and climate from fossil fuel projects, through direct action such as blockades, occupations and street protests.[4] Maps like these can reveal the spatial 'cost-shifting' problem (Kapp 1963) inherent to the long-distance supply chains associated with EU energy security policies. EU energy policy may be increasingly directed towards renewables at the regional level, but the larger picture involves significant investment in and support for fossil energy supply lines (e.g. oil and gas pipelines and LNG import terminals). Inspired by long-standing social movements against fossil fuel extraction, such as the Ogoni People in the Niger Delta and the Yasuni initiative in Ecuador, communities enmeshed in the EU's fossil fuel (and biomass) supply lines are increasingly demanding these fuels remain in the ground. Acts of resistance at the 'sharp end' of energy policy implementation are diverse and widespread. They include, for example, the Ende Gelände mass civil disobedience in Germany, mobilisation against offshore drilling in southern Portugal and pan-European activist networks such as Gastivists and Europe Beyond Coal.[5]

11.3.3 Navigating a Multi-scalar World

The tensions and possibilities associated with different geographical scales of action around energy have been central to the project of closer European integration from its beginning, in the form, for example, of the European Coal and Steel Community (Treaty of Paris, 1951) and the European Atomic Energy Community (1957). The Lisbon Treaty (2007) and EC initiatives like the Third Energy Package (2009) affirm these supranational objectives, although a 'major paradox of EU energy policy (remains) the tension between national sovereignty over the energy sector and a community perspective based on solidarity, cooperation and scale' (Szulecki et al. 2016, p. 548). The European Commission now seeks a 'multilevel' approach to energy and climate governance that includes 'the power of bottom-up action', acknowledging the role of cities and local authorities in building resilience and achieving low-carbon transition.[6] A political ecology perspective affirms the significance of geographical scale but, importantly, reconceptualises its relation to energy policy. Rather than an administrative tension centred on fixed scales (e.g. supranational, national, municipal), political ecology understands scale as the outcome of (contested) social processes. Cross-border energy investment, the connections and disconnections made by energy infrastructure and the alliances and solidarities forged by social movements *create* scales of energy production, consumption and governance.

Thus, political ecology identifies a more complex and fluid scalar world than is represented in most policy analysis. Failure to acknowledge how social processes produce scale—and how prevailing scales express and serve the interests of those actors able to establish and entrench them—can lead to a 'scalar trap' (Brown and Purcell 2005): the assumption that one particular scale is a priori more capable of providing desired outcomes (e.g. encouraging democratic participation, giving voice to marginalised populations, equitably distributing benefits). This is a significant insight, given efforts within the EU to distribute governance 'downward', from international and national to subnational, regional and urban scales. Political ecology research indicates such 'shifts' in governance are often less empowering than they first appear. Rather than giving local communities a voice in formulating and implementing policies, they can entrench decision-making power at a national level while saddling local and regional actors with responsibilities for implementation (Cohen and Bakker 2014). There is some evidence for this around current EU climate change policy following the Paris Agreement, where different roles are assumed for actors at certain scales. For example, national governments 'launch initiatives' and set agendas, while cities and civil society are responsible for implementing emission reductions, planning for and building resilience and finding ways to encourage investment. In this context, the Commission's embrace of 'bottom-up action' can be interpreted as a 'flanking mechanism'—a common phenomenon in the context of neoliberal governance—in which national governments encourage civil society actors to provide services (often services that cushion against the destructive effects of open markets) which might otherwise be provided by government, as a means of reducing government 'interference' and freeing up markets (Castree 2008). Political ecology research suggests more democratic and egalitarian policy outcomes can be achieved if marginalised communities are able to engage in 'scale jumping'—moving outside of scalar hierarchies, circumventing gatekeeping mechanisms and making their voices heard on a broader scale.

11.4 Conclusion

Political ecology is a well-established interdisciplinary Social Science field with a record of work in relation to environmental policy and management. Its orientation towards bringing about emancipatory forms of social and environmental change through the generation of new knowledge builds on a tradition of critical thought and praxis. It is internally diverse, having been shaped by several different intellectual traditions and grounded concerns (e.g. air and water pollution, land dispossession, hazards and risk), although we have drawn out unifying themes in the interests of developing a political ecology perspective on EU energy policy.

Political ecology's critical perspective challenges many of the premises of EU energy policy; its way of working with affected communities—and the value it attributes to their knowledge, concepts and demands—offers an alternative to 'top-down' policy accounts. Implementing a political ecology perspective through research can open up new ways of thinking about the objectives, assumptions and methods of energy policy in the EU: in this sense, it can be a powerful tool in the collective effort to craft sustainable and socially just energy futures. At the same time, political ecology is also alive to how conceptual innovation and new knowledge can also be co-opted to preserve, rather than dissolve existing structures of social power: it is, therefore, always in (creative) tension with the formal apparatus of policy.

We suggest a political ecology perspective on EU energy policy can be pursued simultaneously at several levels. It can involve research with affected communities as outlined above; deconstructing energy policy's objectives, discourses and guiding concepts; or working creatively with frictions and alternative agendas already present in policy, such as the inclusion of demand moderation alongside the older language of energy efficiency in the Commission's Framework Strategy for a Resilient Energy Union (2015). Finally, a political ecology perspective can also require getting closer to the process of energy policy implementation by the Commission and Member States, to understand how social power is reproduced (and how it may be challenged) through institutional, epistemic and market mechanisms.

Acknowledgements We wish to thank three anonymous reviewers, the editors, Magdalena Kuchler and Antti Silvast for helpful comments on the first draft and colleagues from CES at the University of Coimbra who participated in the workshop. We also acknowledge ENTITLE, European Network in Political Ecology (www.politicalecology.eu), an FP-7 funded Initial Training Network under the Marie Curie Actions (contract no. 289374) which initiated the authors' collaboration.

NOTES

1. The potential of this perspective may be glimpsed in the European Environment Agency's *Late Lessons from Early Warnings* reports, on the environmental and public health impacts associated with asbestos, benzene, sulphur dioxide and radiation from Chernobyl and Fukushima. They show how traditional divisions of scientific knowledge and misplaced certainty created a 'recurring nightmare' in which short-term interests triumphed over long-term collective vision (Harremoës et al. 2001; EEA 2013).

2. Among these shortcomings are a blindness to the Jevons paradox, the counter-intuitive way in which gains in efficiency via technological change end up expanding (rather than decreasing) resource consumption (originally noted by British economist William Stanley Jevons in the nineteenth century).
3. See https://ejatlas.org/conflict/trans-adriatic-pipeline-in-puglia-italy.
4. See www.ejatlas.org. The term Blockadia originates in the movement against the Keystone XL pipeline in the US. It was later popularised by Naomi Klein who, in her book *This Changes Everything (2015)*, describes it as the 'roving transnational conflict zone [...] where 'regular' people are stepping in where our leaders are failing'.
5. For details see https://www.ende-gelaende.org/en/; http://www.gastivists.org/; https://beyond-coal.eu/.
6. Commission Communication 2016/110/EC (02 March 2016) *The Road from Paris: assessing the implications of the Paris Agreement and accompanying the proposal for a Council decision on the signing, on behalf of the European Union, of the Paris agreement adopted under the United Nations Framework Convention on Climate Change.*

REFERENCES

Barca, S. (2011). Energy, Property, and the Industrial Revolution Narrative. *Ecological Economics, 70*(7), 1309–1315.

Barca, S., & Delicado, A. (2016). Anti-nuclear Mobilization and Environmentalism in Europe. A View from Portugal, 1976–1986. *Environment and History, 22*(4), 497–520.

Bridge, G., McCarthy, J., & Perreault, T. (2015). Editors' Introduction. In T. Perreault, G. Bridge, & T. McCarthy (Eds.), *The Routledge Handbook of Political Ecology* (pp. 3–18). London: Routledge.

Brown, J. C., & Purcell, M. (2005). There's Nothing Inherent About Scale: Political Ecology, the Local Trap, and the Politics of Development in the Brazilian Amazon. *Geoforum, 36*, 607–624.

Bumpus, A. G., & Liverman, D. (2008). Accumulation by Decarbonization and the Governance of Carbon Offsets. *Economic Geography, 84*, 127–155.

Burke, M. J., & Stephens, J. C. (2017). Energy Democracy: Goals and Policy Instruments for Sociotechnical Transitions. *Energy Research & Social Science, 33*, 35–48.

Caneppele, S., Riccardi, M., & Standridge, P. (2013). Green Energy and Black Economy: Mafia Investments in the Wind Power Sector in Italy. *Crime, Law and Social Change, 59*(3), 319–339.

Castree, N. (2008). Neoliberalising Nature: The Logics of Deregulation and Reregulation. *Environment and Planning A, 40*, 131–152.

Cohen, A., & Bakker, K. (2014). The eco-scalar fix: rescaling environmental governance and the politics of ecological boundaries in Alberta, Canada. *Environment and Planning D: Society and Space, 32*(1), 128–146.

EEA. (2013). *Late Lessons from Early Warnings: Science, Precaution, Innovation*. Copenhagen: European Environment Agency.
Harremoës, P., Gee, D., MacGarvin, M., Stirling, A., Keys, J., Wynne, B., & Guedes Vaz, S. (2001). *Late Lessons from Early Warnings: The Precautionary Principle 1896–2000, Environmental Issue Report* (Vol. 22). Copenhagen: European Environment Agency.
Högselius, P., van der Vleuten, E., & Kaijser, A. (2015). *Europe's Infrastructure Transition: Economy, War, Nature*. London: Springer.
Hornborg, A. (2006). Footprints in the Cotton Fields: The Industrial Revolution as Time–space Appropriation and Environmental Load Displacement. *Ecological Economics, 59*, 74–81.
Huber, M. (2015). Energy and Social Power—From Political Ecology to the Ecology of Politics. In T. Perreault, G. Bridge, & T. McCarthy (Eds.), *The Routledge Handbook of Political Ecology* (pp. 481–492). London: Routledge.
Huber, A., Gorostiza, S., Kotsila, P., Beltrán, M. J., & Armiero, M. (2016). Beyond "Socially Constructed" Disasters: Re-politicizing the Debate on Large Dams Through a Political Ecology of Risk. *Capitalism Nature Socialism, 28*(3), https://doi.org/10.1080/10455752.2016.1225222.
Kander, A., Malanima, P., & Warde, P. (Eds.). (2013). *Power to the People. Energy in Europe Over the Last Five Centuries*. Princeton: Princeton University Press.
Kapp, K. W. (1963). *Social Costs of Business Enterprise* (2nd Enlarged ed.). Bombay and London: Asia Publishing House.
Loftus, A. (2017). Political Ecology I: Where Is Political Ecology?. *Progress in Human Geography*. https://doi.org/10.1177/0309132517734338.
M'Gonigle, R. M. (1999). Ecological Economics and Political Ecology: Towards a Necessary Synthesis. *Ecological Economics, 28*(1), 11–26.
Martinez-Alier, J., Kallis, G., Veuthey, S., Walter, M., & Temper, L. (2010). Social Metabolism, Ecological Distribution Conflicts, and Valuation Languages. *Ecological Economics, 70*(2), 153–158.
Martinez-Alier, J., Anguelovski, I., Bond, P., Del Bene, D., Demaria, F., Gerber, J.-F., Greyl, L., Haas, W., Healy, H., Marín-Burgos, V., Ojo, G., Porto, M., Rijnhout, L., Rodríguez-Labajos, B., Spangenberg, J., Temper, L., Warlenius, R., & Yánez, I. (2014). Between Activism and Science: Grassroots Concepts for Sustainability Coined by Environmental Justice Organizations. *Journal of Political Ecology, 21*, 19–60.
Petridis, P., Muraca, B., & Kallis, G. (2017). Degrowth: Between a Scientific Concept and a Slogan for a Social Movement. In J. Martínez-Alier & R. Muradian (Eds.), *Handbook of Ecological Economics* (pp. 176–200). Cheltenham: Edward Elgar.
Robbins, P. (2011). *Political Ecology: A Critical Introduction*. Hoboken: John Wiley and Sons.
Salleh, A. (Ed.). (2009). *Eco-Sufficiency and Global Justice: Women Write Political Ecology*. London: Pluto Press.

Szulecki, K., Fischer, S., Gullberg, A., & Sartor, O. (2016). Shaping the 'Energy Union': Between National Positions and Governance Innovation in EU Energy and Climate Policy. *Climate Policy, 16*(5), 548–567.

Temper, L., Demaria, F., Scheidel, A., Del Bene, D., & Martinez-Alier, J. (2018). The Global Environmental Justice Atlas (EJAtlas): Ecological Distribution Conflicts as Forces for Sustainability. *Sustainability Science, 13*(3), 573–584.

White, D., Rudy, A., & Gareau, B. (2016). *Environments, Natures and Social Theory. Towards a Critical Hybridity.* London: Palgrave Macmillan.

Open Access This chapter is licensed under the terms of the Creative Commons Attribution 4.0 International License (http://creativecommons.org/licenses/by/4.0/), which permits use, sharing, adaptation, distribution and reproduction in any medium or format, as long as you give appropriate credit to the original author(s) and the source, provide a link to the Creative Commons license and indicate if changes were made.

The images or other third party material in this chapter are included in the chapter's Creative Commons license, unless indicated otherwise in a credit line to the material. If material is not included in the chapter's Creative Commons license and your intended use is not permitted by statutory regulation or exceeds the permitted use, you will need to obtain permission directly from the copyright holder.

Afterword 1: Important Contributions Towards Renewal of a Stubborn Energy Research and Policy Agenda

Harold Wilhite is Professor Emeritus at the Centre for Development and the Environment, University of Oslo. He has numerous publications on sustainable energy use, including his book from 2016 entitled 'The Political Economy of Low Carbon Transformation: Breaking the Habits of Capitalism'. He was the Founding Director of the European Council for an Energy Efficient Economy (eceee) and has participated in numerous international policy efforts on sustainable energy consumption, including those initiated by the United Nations Environmental Programme (UNEP), the Commission on Sustainable Development (CSD) and the Organisation for Economic Co-operation and Development's (OECD) Environment Directorate.

This edited book makes an important contribution to efforts to recast sustainable energy policy in light of the new demands on energy production, delivery and consumption, which are fostered by the need for more equitable energy access and for rapid reduction of energy use due to climate change. Virtually every chapter asserts that an integration of perspectives from differing academic disciplines and across policy fields will be needed to address the challenges looming ahead. Particular emphasis is given to the need to draw the Social Sciences and Humanities (SSH) into the efforts to theorise, research and frame policies that will move Europe towards low-energy and low-carbon societies.

In the following discussion, I organise and discuss the core chapters under three headings: the theoretical framing of energy delivery and use, research designs accommodating SSH into an interdisciplinary approach and new topics that are important, but are under-theorised and under-researched, in the energy domain.

The Theoretical Framing of Energy Delivery and Use

Energy efficiency research and policy of the past four decades have been stubbornly resistant to absorbing new theoretical approaches to production, provision and consumption. This is in spite of a steady stream of evidence arguing for the dominant technical, economic and (more recently) 'behavioural' agendas to be refreshed. Nevertheless, there does seem to finally be grudging acceptance of the need for a new 'integrative' approach to research and policy formulation, but this needs fostering. The contributions to this book do just that. The Silvast et al. chapter (Chap. 7) addresses the concept of energy integration and the need for including SSH research in the theoretical framing of energy systems, which are today fragmented both in theory and in practice. The chapter adds to a growing critique of today's dominating theoretical approaches to energy sustainability, which deny the agency of historical experience, both individual and collective, as well as social relations and cultural situatedness in structuring practices. The interdisciplinary team of authors make a solid argument for the need for greater attention to perspectives from History, Political Science, Sociology, Anthropology and Science and Technology Studies. The Bridge et al. chapter (Chap. 11) makes a case for bringing perspectives from Political Ecology to the analysis of energy systems. The authors give much-needed attention to the importance of a largely ignored factor in determining energy service access and consumption: the power that derives from social and economic hierarchies. The Genus et al. chapter (Chap. 9) examines how the integration of SSH has been addressed in concrete EU research programmes, arguing that despite efforts to broaden and integrate SSH perspectives, programmes remain largely technically focused. They emphasise that bottom-up, qualitative research is sorely absent from efforts to theorise the needed energy transition—a point that runs through other chapters too. This places the chapters in this book on the cutting edge of a new effort in energy research and policy to integrate the results of qualitative research into the analysis of new policy pathways.

Research Designs Integrating SSH Research into an Interdisciplinary Approach

While there is grudging acknowledgement in energy research and policy that an interdisciplinary and integrative approach is essential to a more robust agenda, little progress has been made in finding ways to actually make it happen in practice. Hiteva et al. (Chap. 8) address how this could be accomplished in energy modelling, arguing for the need for synergy between two central modelling approaches used in energy efficiency research and policy (technical and agent-based) and addressing how the absence in predominant models of social contextualisation could be compensated. The authors propose that dominating top-down models would benefit from the integration of the results of ethnographic research, though the concrete steps to accomplish this are not fully developed. The Higginson et al. chapter (Chap. 5) addresses this question of how to merge and analyse findings from qualitative and quantitative research. This is an important chapter because it addresses a challenge confronted by the increasing number of research projects taking on interdisciplinary methods and struggling with how to achieve 'data synergy'. It addresses theoretical, methodological and practical issues involved in coordinating research and analysis using qualitative and quantitative data sources. Based on my own work—particularly as leader of a long-term, strategic programme at the University of Oslo aimed at fostering interdisciplinary research on energy sustainability—this chapter fills a gap in our understanding of how to go about concrete interdisciplinary research collaboration. This chapter includes lots of detail on problems with coordination of research instruments and provides insights on lessons learned, making an accessible and important contribution to the development of interdisciplinary methods.

New Topics Important to Energy Research and Policy

Many of the chapters emphasise the importance of drawing the life experiences of non-energy experts into our understanding of energy use. An example of this is Åberg et al.'s discussion (Chap. 4), which draws on the narratives of women from differing cultural contexts in order to identify themes for future research and policy. Gender is finally being given the attention it deserves in energy studies, and the chapter's focus on women's perspectives contributes to deeper understanding of its relevance. In

Middlemiss et al. (Chap. 2), the authors also contribute insights on the questions surrounding how to deal with energy poverty, emphasising the importance of researching lived experience through an interdisciplinary research design. Kerr et al. (Chap. 3) take on the issue of scales, arguing that community engagement is essential to the renewable harvesting of marine resources. This supports my own assessment that while national energy agendas tend to be steeped in inertia, a huge potential exists in bottom-up, community-driven initiatives (Geels 2014; Wilhite 2013). The McCarthy et al. chapter (Chap. 6) supports the important point that the ways that people use energy are not just a property of individuals but are strongly influenced by laws, regulations and norms (Shove et al. 2013; Sahakian and Wilhite 2014). Their discussion is a much-needed exploration of how legal instruments can be framed and used to encourage energy conservation in Multi-owned Properties (MoPs). Both MoPs and the role of legal instruments are under-researched and deserve more attention in future research and policy. The Turnheim et al. chapter (Chap. 10) takes up another largely neglected topic associated with the coming energy transition: the societal tensions that are likely to arise and the need to anticipate and address them through an interdisciplinary design.

As a concluding thought, this book represents a refreshing step forward towards a renewal of energy research and policy strategies encompassing theory, interdisciplinary integration and methodologies. It constitutes a rich source of ideas and experiences for a transformative research and policy agenda.

Harold Wilhite, Centre for Development and the Environment, University of Oslo; European Council for an Energy Efficient Economy (eceee).

References

Geels, F. W. (2014). Regime Resistance Against Low-carbon Transitions: Introducing Politics and Power into the Multi-level Perspective. *Theory, Culture and Society, 31*(5), 2–40.

Sahakian, M., & Wilhite, H. (2014). Making Practice Theory Practicable: Towards More Sustainable Forms of Consumption. *Journal of Consumer Culture, 14*(1), 25–44.

Shove, E., Pantzar, M., & Watson, M. (2013). *The Dynamics of Social Practice: Everyday Life and How It Changes*. London: Sage.

Wilhite, H. (2013). Energy Consumption as Cultural Practice: Implications for the Theory and Policy of Sustainable Energy Use. In S. Strauss, S. Rupp, & T. Love (Eds.), *Cultures of Energy*. San Francisco: Left Coast Press.

Afterword 2: A New Energy Storyline

Inês Campos is the coordinator of Horizon 2020 energy-related Social Sciences and Humanities project, PROSEU (PROSumers for the Energy Union: mainstreaming active participation of citizens in the energy transition). As a social scientist with a PhD in Climate Change and Sustainable Development Policies, she has developed transdisciplinary research on climate change mitigation and adaptation in Europe and coordinated policy planning pilot cases.

This brief Afterword navigates through this collection on *Advancing Energy Policy* to reflect on some of the key ideas raised for future research and how they relate to a potential new energy transition storyline centred on the role of citizens.

The importance of Social Sciences and Humanities (SSH) in the 'imaginaries' of energy research and policy institutions—that is, 'visions of desirable futures'; see definition in Genus et al. (Chap. 9; p. 133)—has been discussed. Several chapters argued for the importance of transdisciplinary research and how this can bring to the foreground the voices of those whose living conditions are (positively or negatively) affected by new (energy-related) material cultures (e.g. coastal communities in the case of marine renewable energy) (Kerr et al., Chap. 3). In this way, the chapters make the case for a transformation in the way we do and perceive science and research (Weber and Rohracher, 2012) that keeps up with the ongoing energy transition. Such transformation is characterised by certain

© The Author(s) 2018
C. Foulds, R. Robison (eds.), *Advancing Energy Policy*,
https://doi.org/10.1007/978-3-319-99097-2

words the authors use, such as 'inclusiveness', 'integrated', 'holistic', 'narratives', 'bottom-up', 'participatory', and 'ethnographic' approaches, seen as required to grasp the complexities of the energy system, and implying that the integration of SSH into energy research so far is still not living up to these words. This 'new' energy research starts from how questions are framed and is equally crucial in the process of data collection (Higginson et al. in this collection—Chap. 5).

To pull out some examples, chapters argued for the need to integrate a more comprehensive interpretation of social practices in building governance (McCarthy et al., Chap. 6), for a response to the call for a 'holistic' approach to deal with energy poverty (Middlemiss et al., Chap. 2), and for an energy transition that is 'democratic'. The chapter on political ecology (Bridge et al., Chap. 11) calls explicitly for a transformation of research practices, from questions framed by reference to the disciplinary traditions (which are likely to provide only a fragmentary understanding of the problem and its solutions) to questions framed by the structure of the problem. This problem structuring approach (Hisschemöller and Hoppe 2017) is crucial for future SSH research and implies the articulation and integration of different sources of information to overcome the pitfalls of complex wicked problems (Buchanan 1992). This applies alike to inclusion of the STEM sciences—for example, the combining of techno-economic models with ethnographic approaches are explored to provide wide-ranging insights for energy-related policies (Hiteva et al., Chap. 8).

'Problem structuring' is the first step in transition management (Loorbach 2010) but has been often overlooked in transition studies more widely. The approach implies a socio-technical systems perspective (Geels 2004), bringing to the foreground the full scope of the 'seamless web' (Silvast et al., Chap. 7; p. 101) of the energy system, wherein people's everyday lives are inseparable from the system's techno-economic and ecological components. Problem structuring relies on the cross-fertilisation of scientific and practical knowledge (Lang et al. 2012). The chapters seem to revolve around an idea that there is a co-evolving and interdependent relation between who we are, our environment (Åberg et al., Chap. 4), and the technical and material world we co-construct; many of the chapters see a need for this to be more recognised in research and policy. Yet this requires an action-research approach, such as problem structuring, capable of addressing the complexity of the issues we are dealing with (Campos et al. 2016).

These chapters offer lessons for other energy-SSH projects, and I'm now going to outline these in the case of the H2020 project I coordinate, PROSEU—*PROSumers for the Energy Union: mainstreaming active participation of citizens in the energy transition*.[1] PROSEU's research is grounded in a problem structuring approach focussed on renewable energy prosumers—these are individuals or collectives who both produce and consume energy from renewable sources.

Considering the possible pathways ahead, and the rising discourse around energy poverty and energy democracy, Europe could be at a crossroads for the future of the energy transition. One scenario is predominantly driven by citizens. In this scenario, the transition continues towards a decentralised system, and the potential losers are the incumbent energy actors (e.g. large utility companies). In another scenario, the (sociotechnical) innovation potential of prosumerism is taken over by incumbent actors, and the transition process is dominated by the latter.

As the transition reaches its 'acceleration phase' (Turnheim et al., Chap. 10), a question emerges: could researchers play the role of 'system integrators' (Silvast et al., Chap. 7), by actively influencing long-term political projects through encouraging a problem structuring and participatory approach to research? PROSEU's interdisciplinary and transdisciplinary research could provide some answers. PROSEU will seek to understand how prosumers are solving their own problems and taking advantage of opportunities, while identifying the key regulatory, economic, and technological (dis)incentive structures for prosumers, leading up to designing a roadmap for the mainstreaming of citizens in the energy transition. The idea from political ecology of energy becoming a 'commons' to be self-produced and shared rather than 'secured and commodified' (Bridge et al., Chap. 11; p. 168) is an important topic for research on renewable energy prosumers. A bottom-up approach is embedded in the PROSEU's 'Living Labs' research (Evans and Karvonen 2011).

When considering the implications of focussing on citizens, other questions are prompted by this book. Could putting energy production/consumption in the hands of local communities help tackle energy poverty? Could local governments play an important role in incentivising prosumerism? What about the potential of the 'Multi-owned Properties' (McCarthy et al., Chap. 6) for the emergence of a new type of prosumerism, centred on neighbourhoods? The finding that 'community mythologies' can influence how technologies are received (Kerr et al., Chap. 3) encourages an investigation that focuses on the socio-political and socio-cultural incentives for renewable energy prosumers. Additionally, establishing a national and transnational space for sharing 'stories' between communities can create a common and shared discourse, a narrative of change.

Inspired by the chapters' accounts of ethnographic research, future studies on renewable energy prosumers, spanning from local communities to businesses or local governments, can provide a lived knowledge of the experience of becoming a prosumer that goes beyond economic, regulatory, or technological issues to touch upon the cultural, everyday life practices and the underlying narratives that can culturally bind diverse prosumers across Europe in a shared collective understanding of the importance of their role at a stage of accelerating transition.

The multiplication of renewable energy prosumers and energy communities across Europe is one example of a new regime, which can change the rules of the game: in this case from centralised to decentralised, from global sources to local sources. In the end, a different storyline for people emerges, which is deep-seated in this book's key ideas for a transformative research and policy related to the energy transition.

Inês Campos, CCIAM group of the Centre for Ecology, Evolution and Environmental Changes of the Faculty of Sciences, University of Lisbon.

Notes

1. PROSEU is one of only two energy-related Social Sciences and Humanities Horizon 2020 projects that were funded in 2017. More information about the project can be found via www.proseu.eu.

References

Buchanan, R. (1992). Wicked Problems in Design Thinking. *Design Issues, 8*(2), 5–21.

Campos, I., Alves, F., Dinis, J., Truninger, M., & Penha-Lopes, G. (2016). Climate Adaptation, Transitions, and Socially Innovative Action-Research Approaches. *Ecology and Society, 21*(1), 13.

Evans, J., & Karvonen, A. (2011). Living Laboratories for Sustainability: Exploring the Politics and Epistemology of Urban Transition. In H. Bulkeley, V. Castán Broto, M. Hodson, & S. Marvin. (Eds.), *Cities and Low Carbon Transitions*, (pp. 126–141). Abingdon and New York: Routledge.

Geels, F. W. (2004). From Sectoral Systems of Innovation to Socio-technical Systems: Insights About Dynamics and Change from Sociology and Institutional Theory. *Research Policy, 33*(6–7), 897–920.

Hisschemöller M., & Hoppe, R. (2017). Coping with Intractable Controversies: The Case for Problem Structuring in Policy Design and Analysis 1. In M. Hisschemôller, R. Hoppe, W. N. Dunn, & J. R. Ravetz (Eds.), *Knowledge, Power, and Participation in Environmental Policy Analysis*, (pp. 47–72). Abingdon and New York: Routledge.

Lang, D. J., Wiek, A., Bergmann, M., Stauffacher, M., Martens, P., Moll, P., Swilling, M. & Thomas, C. J. (2012). Transdisciplinary Research in Sustainability Science: Practice, Principles, and Challenges. *Sustainability Science, 7*(1), 25–43.

Loorbach, D. (2010). Transition Management for Sustainable Development: A Prescriptive, Complexity-Based Governance Framework. *Governance, 23*(1), 161–183.

Weber, K. M., & Rohracher, H. (2012). Legitimizing Research, Technology and Innovation Policies for Transformative Change: Combining Insights from Innovation Systems and Multi-level Perspective in a Comprehensive 'Failures' Framework. *Research Policy, 41*(6), 1037–1047.

Index[1]

A
Accelerated diffusion, 147–155, 157, 158
Action Agenda for European Universities, viii
Action research, 184
Actor Network Theory (ANT), 91
Advocacy, 24, 25
Affordable access/affordability, vii, 167
Agent Based Modelling (ABM), 114, 118–123
Anthropology, 101
Anticipation, 147–158
Apartments, 84, 87, 89
Appliances, 66, 67, 71, 73, 74, 76
Assessment, 158

B
Barrier model, 85
Behaviour, 178
Behavioural change, 137

Best-practice, 22, 24
Big data, 113, 114, 122
Blue economy, 35
Blue growth, 33–44
Bottom-up, 184, 185
Bridger organisations, 42, 43
Building Energy Model (BEM), 66
Bulgaria, 49–53, 56
Business, v, vi
Business models, 136
Bystander Intervention Model, 88

C
Centre for Energy Systems Integration, 99
Citizens, v, vi
Civic epistemologies, 134, 137, 140–141
Climate change, 132, 137
Climate justice, 166
Closures, 134, 137, 140

[1] Note: Page numbers followed by 'n' refer to notes.

Co-creation (of knowledge), 138, 140, 142
Collective action, 90, 91
Collective Efficacy Theory, 88
Colonisation, 50, 54, 55
Common rights, 34
Communities, 166, 168–172
Community (groups), 138
Complementarity, 124
Condominiums, 84, 88, 91
Connectivity, 67, 69–75
Consumer, 168–170
Consumer choice, 137
Controversies, 133, 134, 137, 140
Customer impact assessment, 66–67, 74

D
Data
  collection, 64–66, 69–71
  protocol, 69, 78
  quality, 64, 71, 75
  synergy, 64–78, 78n3
Decarbonisation, 167, 168
Decentralised system, 185
Decision making, v
Degrowth, 166, 168
Demand, energy, 135, 138, 140, 141
Demand Response (DR), 67, 71, 73, 74, 78n7
Dematerialisation, 167
Denmark, 36, 38
Design, 33, 38–40, 43
Diffusion, 147, 149, 153–157
Diffusion phase, 147, 148, 153
Disciplinary chauvinism, 132
Dominance, 164, 165

E
Ecological debt, 166
Ecological modernisation, 167–168
Eco-sufficiency, 168
Education, vii, 17, 21, 22
Edwards, Paul N., 99, 105
Emergence phase, 148
Energy
  consumers, 103
  consumption, 112–124
  demand, 112–124, 135, 138, 140, 141
  democracy, 169
  efficiency, 84–86, 88, 135, 137, 139, 140, 142
  governance, 135, 140
  infrastructure, 166, 169, 170
  justice, 23, 25
  policy, 16–27, 98, 99, 106, 107, 164–166, 168–170, 172
  poverty, 16–27, 52, 53, 55
  research, 132, 133, 141–142
  retrofit, 88, 90–92
  services, 112–124
  transition, v, 165, 167
  use, 138–140
  vulnerability, 21
Energy Roadmap 2050, 48–50, 54
Energy Systems Integration (ESI), 98–108
Energy Union, 132–142, 166
Environmental history, 167
Environmental impacts, 149, 152, 155, 156
Environmental justice, 166, 169
Ethnographic research, 113, 114, 119–123
Ethnography, 184, 186
European Commission, v, vi, 2, 5–6, 26, 48, 54, 56, 57n1, 57n3, 98, 99, 106, 132, 137, 147, 170
European Technology and Innovation Platforms (ETIPs), vi
European Union (EU), 2, 4, 6, 7, 9n2, 16, 17, 25, 33, 47–57, 102, 105–107, 112, 114–116, 131–142, 147, 163–173, 178

European Union Energy Poverty Observatory (EPOV), 16, 17
Everyday life, 103
Evidence/Evidence-based policy, x, 2–4, 6, 113, 114, 121, 123
Expertise, 165, 166

F
Formative phase, 150–154
Fossil fuels, 167
Framing, 133, 137, 139
Funding, 5, 6, 8, 9n5, 10n8

G
Gender, 165, 167
Governance, 84–94, 148, 149, 153–158
Governance of energy, 135, 140

H
Hecht, Gabrielle, 99, 101
History, 98, 102
Horizon 2020 (framework programme), 5, 6, 9n3, 9n5, 10n8, 132–135, 137, 139–142
Hughes, Thomas P., 99, 101, 102
Humanities, 135, 136

I
Imaginaries, 132–142, 183
Income, 18, 19, 21, 22, 26
Incumbents, 148, 150–152, 157, 185
India, 49, 50, 53, 55, 56
Indicators, 18, 25, 26
Infrastructures, 101–105, 107
Innovation, 136, 137, 140, 142, 148, 149, 151, 152, 156
  technical, 140
Institutionalisation, 132–134, 141

Integration, 4–6, 8, 132–142, 177–180
Interdisciplinary, 4–8, 83–94, 99–102, 123, 124, 135, 138, 142, 148, 164, 166, 171
  bridging, 158
  methods, 179
  research, v–vi, 134
International Institute for Energy Systems Integration (iiESI), 98
International Network for Social Studies of Marine Energy (ISSMER), 33
Internet of Things (IoT), 71

J
Jevons paradox, 173n2
Joint Research Centre (JRC), 2

K
Knowledge, 164–166, 168, 171, 172, 172n1
  co-creation of, 138, 140, 142
  co-inquiry, 140
  exchange, 138
  generation, 138
  sharing, 140

L
Labour, 167
Land-grabbing, 51, 168
Life experiences, 179
Lived experience, 16, 28n1
Living lab, 138, 140–142
Low-carbon transitions, 147–158

M
Mainstreaming, 6, 10n7
Marginalisation, 165

Marine renewable energy (MRE), 43
Modern economic growth, 167–168
Monitoring/evaluation, 25, 64, 66, 70, 74, 75, 77
Multidisciplinary (research), vii–viii, 16, 136
Multi-owned Properties (MoPs), 84–93
Multi-stakeholder, 6
Mythologies, 33, 36–37, 40, 42, 43

N
the Netherlands, 16–19, 22, 24, 26
New entrants, 150
North Atlantic, 33
Nudge (approach to policy), 89, 90, 139

O
Observation, 119–123
Optimisation, 117, 119
Orkney, 38
Ownership, 34–37, 40, 43

P
Partnerships, vii–viii, 24–25
Path dependence, 156
Platform, 5, 9, 10n8
Policy goals, 2
Policy integration, 132–134, 140, 141
Policymaker, 112–118, 120–124
Policymaking, 112–114, 117, 120, 121, 123, 124
Political ecology, 164–172
Political science, 98
Political struggle, 154
Poverty, 16, 22
Practice culture(s), 137, 140, 142
Practice(s), 132–142

social, 135
Praxis, 165, 171
Problem structuring approach, 184, 185
Property law, 85, 86
Prosperity, ix
Prosumer, 185, 186

Q
Qualitative, 17, 26, 27, 28n1, 40, 41, 65, 67, 74, 78, 119, 122, 134, 135, 140, 141, 157, 158
Quantitative, 26, 40, 41, 67, 69, 71, 78, 119, 122, 134, 155

R
RealValue, 65, 71, 73
Renewable Energy Technology (RET, RETs), 147–158
Representation, 22
Research & Innovation, 5
Resilience, 155–157
Rich data, 122
Risk, 164, 168, 171

S
Sami, 49–51, 54, 57n1
Sápmi, 50, 51
Scale, 164–166, 170, 171
Scale as a method, 98, 99, 105–108
Scaling, 164–166, 170, 171
Scarcity, 164, 165, 167
Science and Technology Studies (STS), 99, 101
Science, Technology, Engineering, and Mathematics (STEM), 3–5, 132, 134
Scotland, 33, 35, 38
Seamless web, 99, 101–103, 107

Security, 164, 167, 169
Shadowing, 119–123
SHAPE ENERGY, 5–6, 9
Silos, ix–x
Simulation, 117, 119
Smart Electric Thermal Storage (SETS), 66, 69, 75, 76
Social Identity Theory, 88, 91
Social innovation, 136
Social licence to operate (SLO), 36
Social marketing, 90
Social metabolism, 165
Social movements, 164, 165, 169, 170
Social policy/social learning, 23
Social power, 164, 172
Social practice, 135
Social science(s), 132–142
Social Sciences and Humanities (SSH), 33, 34, 36–38, 40–44, 132–137, 139–142
Socio-ecological approaches, 149, 155
Socio-ecological systems, 154, 155
Sociology, 101
Socio-political, 166
Socio-technical, 64
  approaches, 148, 149
  change, 149
  imaginaries, 133, 134, 137
  niches, 153
  regimes, 154
  systems, 99, 101, 102, 149, 154
Spain, 16–20
SSH-flagged, 5
STEM, *see* Science, Technology, Engineering and Mathematics
Stories/narrative, 48, 51, 54, 55, 57, 184–186
Strategic Energy Technology Plan (SET-Plan), vi
Sustainability transitions, 157
Sustainable energy, 166–168
Sweden, 49–51

Synchronisation, 69–70
Systemic approach, x

T
Technical innovation, 140
Techno-economic, 57
  approaches, 149, 155
  models, 114, 115, 117–123
  systems, 149
Technopolitics, 99, 101
Theory of Planned Behaviour (TPB), 88
Theory/theoretical framing, 178, 180
Thick data, 113, 122, 123
Time, 64, 66, 69, 71, 77
Transdisciplinary, 183, 185
Transformation, 148, 155, 158, 183, 184
Transnational networks, 25
Trial(s)/field trial(s), 64–66

U
Understanding, 147–158
United Kingdom (UK), 16, 17, 20–21, 24, 25
United Nations Framework Convention, ix
Universities, vii–ix
Universities in the SET-Plan (UNI-SET), vii

V
Vanua, 34

W
Wind energy, 36, 39
World Bank, 52, 55

The manufacturer's authorised representative in the EU is Springer Nature Customer Service Centre GmbH, Europaplatz 3, 69115 Heidelberg, Germany. If you have any concerns regarding our products, please contact ProductSafety@springernature.com

Printed and bound by CPI Group (UK) Ltd, Croydon, CR0 4YY

23/03/2026

02076670-0002